Lecture Notes in Biomathematics 93

Managing Editor:
S. A. Levin

Editorial Board:
Ch. DeLisi, M. Feldman, J. B. Keller, M. Kimura
R. May, J. D. Murray, G. F. Oster, A. S. Perelson
L. A. Segel

W0051430

Ziad Taïb

Branching Processes and Neutral Evolution

Springer-Verlag Berlin Heidelberg GmbH

Author

Ziad Taïb
Department of Mathematics
Chalmers University of Technology
and the University of Göteborg
S-41296 Göteborg

Mathematics Subject Classification (1991): 60-02, 60J80, 92-02, 92A10, 92A12, 92A15

ISBN 978-3-540-55529-2   ISBN 978-3-642-51536-1 (eBook)
DOI 10.1007/978-3-642-51536-1

© Springer-Verlag Berlin Heidelberg 1992
Originally published by Springer-Verlag Berlin Heidelberg New York in 1992

Typesetting: Camera ready by author/editor
46/3140-543210 - Printed on acid-free paper

PREFACE

The Galton-Watson branching process has its roots in the problem of extinction of family names which was given a precise formulation by F. Galton as problem 4001 in the Educational Times (17, 1873). In 1875, an attempt to solve this problem was made by H. W. Watson but as it turned out, his conclusion was incorrect. Half a century later, R.A. Fisher made use of the Galton-Watson process to determine the extinction probability of the progeny of a mutant gene. However, it was J.B.S. Haldane who finally gave the first sketch of the correct conclusion. J.B.S. Haldane also predicted that mathematical genetics might some day develop into a "respectable branch of applied mathematics" (quoted in M. Kimura & T. Ohta, Theoretical Aspects of Population Genetics. Princeton, 1971). Since the time of Fisher and Haldane, the two fields of branching processes and mathematical genetics have attained a high degree of sophistication but in different directions.

This monograph is a first attempt to apply the current state of knowledge concerning single-type branching processes to a particular area of mathematical genetics: neutral evolution. The reader is assumed to be familiar with some of the concepts of probability theory, but no particular knowledge of branching processes is required. Following the advice of an anonymous referee, I have enlarged my original version of the introduction (Chapter Zero) in order to make it accessible to a larger audience.

Göteborg, Sweden, November 1991.   Ziad Taib

ACKNOWLEDGEMENTS

First of all I want to express my gratitude to O. Nerman and P. Jagers, my colleagues and friends. Many thanks are also due to R. Doney, an anonymous referee for many helpful comments and to Yumi Karlsson for taking good care of my manuscript. This work has been supported by the Swedish National Science Research Council.

CONTENTS

INTRODUCTION

The last two decades have witnessed an increasing interest in models for the behaviour of what we shall call labelled populations. If we restrict ourselves to asexual populations, then a labelled population can be described as one where every individual carries a label which she does or does not transmit to her offspring. An individual not receiving her mother's label is assumed to carry a completely new label neither currently nor previously encountered in the population. Such an individual is called a mutant. Individuals carrying distinct labels are assumed to behave in the same way, at least with respect to reproduction and viability. This provides models for the study of the genetical structure of haploid populations subject to neutral mutations at some single locus under the infinite alleles hypothesis of population genetics. A first model of this kind can be traced back, in an embryonic form, to [Kimura & Crow, 1964]. Since then the topic has been subject to intensive research.

The purpose of this monograph is to provide a theory for the situation arising when the underlying population obeys the laws of a general (or Crump-Mode-Jagers) single-type branching process. It has thus two ingredients: branching processes and the neutral theory of molecular evolution. An excellent review of the neutral theory can be found in [Kimura, 1983]. A detailed (technical) description of the general branching process is provided in Chapter One. Throughout the book, we use less specific terminology and talk of labels instead of alleles.

The plan of this introduction is as follows. Its first section contains an attempt to give an intuitive account of the branching process approach. In § 0.2, the neutral theory is discussed (very rapidly). The last section is devoted to a summary of the results of the book.

0.1. THE GENERAL BRANCHING PROCESS

A general branching process is, following [Jagers & Nerman, 1984a], a model for the description of the evolution of asexual populations in continuous time. In the most general setting, every individual in such a process inherits, at birth, a genotype $\gamma \in \Gamma$, Γ being the set of all possible genotypes. The most important feature of a general branching process is the concept of life history (or life for short). In many population genetics models, the individuals of the next generation are produced by a random drawing of gametes from a gene pool. In a general branching process, the life histories are chosen in a similar manner. The set, Ω, of all possible life histories may be viewed as a life history pool from which an individual of type γ chooses a life history, $\omega \in \Omega$, according to some random mechanism which we describe using a probability law, P_γ , to be referred to as the life law of individuals of type γ. Doing so implies that two individuals of the same genotype will choose their life histories using the same random mechanism. Our main interest will, however, be in the so called single–type case where $P_\gamma = P$ for all $\gamma \in \Gamma$.

At this early stage, we will choose to refrain from specifying the life history concept too much. One should think of a life history as a complete description of the destiny of an individual. If we, for example, let λ stand for the life span of an individual, then λ will be a real valued (measurable) function of the life history, ω , of this individual; and once the random element ω is chosen, $\lambda(\omega)$ will be a fixed life length. In the terminology of probability theory, Ω is the so called outcome space and λ is a random variable defined on Ω. In order to complete this picture, we let $\mathscr{A}$ be a σ–algebra in the set Ω (i.e. a class of permissible events in Ω) and take the triplet $(\Omega, \mathscr{A}, P)$ as our basic probability space at the individual level. The reproduction of an individual having ω as life history can be described through a sequence, $(\tau(k, \omega))_{k \in N}$, $N = \{1, 2, ...\}$, of random variables to be interpreted as the successive reproduction ages (i.e. $\tau(k, \omega)$ is the age at which she begets her kth child). A useful way of summarizing the actual reproduction of an individual with life history ω is to let $\tau(k, \omega) = \infty$ whenever $\tau(k, \omega) > \lambda(\omega)$ and to define

$$\xi(A, \omega) = \#\{k; \tau(k, \omega) \in A \text{ and } \tau(k, \omega) < \infty\}$$

where A is some (measurable) subset of the real line. If A is an interval of the type $[0, t]$, then $\xi([0, t], \omega)$ will stand for the number of children born to an individual with life history ω during the age interval $[0, t]$. For ease of notation the argument ω will be omitted, i.e. we will write λ instead for $\lambda(\omega)$, $\xi(A)$ instead for $\xi(A, \omega)$ etc. We will also write $\xi(t)$ instead for $\xi([0, t])$. In what follows, ξ will be called the reproduction point process. Formally, point processes are integer-valued random measures, but for us they can be viewed as general schemes for scattering points on the line. Familiar examples of point processes are the (time-homogeneous) Poisson process and renewal processes. The generality of the general branching process comes partly from the fact that it allows reproduction according to an arbitrary point process. A trivial (but important) example is the case where all reproduction is concentrated at age 1. In this case, we obtain the generation counting Galton-Watson branching process, the simplest of all branching processes. One of the most basic parameters of the general branching proccess is

$$\mu(t) = E[\xi(t)]$$

where E denotes expectation with respect to the life history law P. It is possible to show that μ defines a measure on the line.

Having described the individual behaviour, let us now describe how things work at the level of the whole population. We do this by assuming complete independence between the choices of life histories of distinct individuals. If I denotes the set of all possible population can be defined on the product space $(\Omega^I, \mathscr{A}^I, P^I)$ which we take as our basic probability space. According to this way of describing our population process, the basic outcome space is $\Omega^I = \underset{x \in I}{\Pi} \Omega$ with elements $(\omega_x)_{x \in I}$ containing one life history coordinate, ω_x, for every possible individual x in the population. The product form of P^I reflects the above assumption about the life histories of distinct individuals being independent and identically distributed (i.i.d). For notational convenience we will write P instead of P^I.

Most of the results in this book are obtained using the idea of associating random characteristics, χ, with the individuals. Such random characteristics can be thought of as weights or scores which can be varied to yield different ways of counting or measuring the population. In its simplest form, a random characteristic is a function of the life history and the age of one single individual (the point here being that the random characteristic associated with an individual does not depend on the life histories of other individuals). A suitable name for this type of random characteristics is individual characteristics. In order to be able to tell more about random characteristics, we have to describe how the population evolves in real time.

Imagine a branching population starting at 0 with one single ancestor of age zero. It is, in principle, possible to let the population start with any finite number of ancestors having arbitrary ages, but this will only render our heavy notation heavier. The assumed independence between individuals will in any case imply that a population initiated by k distinct ancestors will consist of k independent populations. The birth moment, σ_0, of our single ancestor is thus, by definition, equal to zero. We will further let σ_x denote x's birth moment, x being an arbitrary individual. Such birth moments are completely specified by the recursion

$$\sigma_{xk} = \sigma_x + \tau(k, \omega_x)$$

where xk stands for x's kth child and ω_x is x's life history.

Define the branching process counted by the random characteristic χ as

$$Z_t^\chi = \underset{x \in I}{\Sigma} \chi_x(t - \sigma_x)$$

and assume for the time being that χ is individual and that it vanishes for negative arguments (i.e. equals zero whenever $t - \sigma_x > 0$ fails, $t - \sigma_x$ being x's age at time t). The simplest random characteristic is obtained when

EX. 1 $\chi_x(t - \sigma_x) = 1_{R_+}(t - \sigma_x).$

In this case Z_t^χ counts all individuals born at or before t, both dead and alive. We will agree to call this particular process the total population process and will denote it by y_t.

Another important special case is

EX. 2 $\qquad \chi_x(t-\sigma_x) = 1_{[0,\lambda_x)}(t-\sigma_x)$

which yields the process Z_t counting all individuals still alive at time t.

However, the real the strength of the random characteristic idea comes from the fact that $\chi_x(t-\sigma_x)$ may also depend on x's whole progeny. Random characteristics can in fact be even more general than this as will be seen later. Our main interest will be in processes meeting the following technical requirements:

(1) $\qquad \mu(\infty) = E[\xi(\infty)] > 1.$

In words, $\mu(\infty)$ stands for the mean number of children ever born to an individual. Processes satisfying this condition are called supercritical and are known to escape extinction with positive probability.

(2) $\quad \exists \; \alpha \in R$ such that $\hat{\mu}(\alpha) = \int_0^\infty e^{-\alpha u} \mu(du) = 1.$

The notation $\hat{f}(\alpha)$ will be used to denote the Laplace transform $\int_0^\infty \alpha e^{-\alpha u} f(u)du$ of a function f. α is the so called Malthusian parameter which describes the exponential growth of the population (cf. below).

(3) $\qquad \beta = \int_0^\infty u e^{-\alpha u} \mu(du) < \infty.$

(4) $\quad \mu$ (as a measure) is non-lattice,

i.e. it cannot be supported by any lattice $\{0, d, 2d,...\}$, $d > 0$, as will be the case if all reproduction is concentrated at equally spaced, predetermined ages. The lattice case is discussed in Chapter Three (cf. also APPENDIX B).

In view of biological applications, it is certainly realistic to assume that the variance of the number of children born to an individual is finite or equivalently

(5) $\qquad Var[\hat{\xi}(\alpha)] < \infty.$

But it turns out that in many cases we can afford being less demanding and require

(5') $\qquad E[\hat{\xi}(\alpha) \log^+ \hat{\xi}(\alpha)] < \infty.$

In some cases we will also need the following condition

(6) $\qquad \exists \; \alpha' < \alpha$ such that $\hat{\mu}(\alpha') < \infty.$

(The exact relation between this condition and condition (5) is rather complicated. It can be shown by counterexamples that neither of these implies the other one).

What makes branching processes counted by random characteristics very suitable in view of biological modelling is the fact that their asymptotic behaviour can be caught through a number of basic convergence theorems. Using those convergence results does not require more than specifying μ, calculating α and β and specifying χ. For all "nice" χ:s one then immediately has results of the form (cf. APPENDIX A for the exact formulations)

(a) $\qquad e^{-\alpha t}[Z_t^\chi] \longrightarrow E[\hat\chi(\alpha)] / \alpha\beta.$

(b) $\qquad e^{-\alpha t}Z_t^\chi \longrightarrow E[\hat\chi(\alpha)]W_\infty / \alpha\beta,$

$\qquad$ in probability on the (Ω^I-) set $\{y_t \longrightarrow \infty\}$ or equivalently

$\qquad$ on the set where the population escapes extinction.

From (a) and (b), we see that the expectation of such a process will grow exponentially as $e^{\alpha t}$ times a constant which depends on the way of counting the population. The process itself has the same limit multiplied by the usual martingale limit random variable, W_∞, of branching processes. At this stage, the exact nature of this limit variable is not important. As will be seen below, the following ratio limit theorem is much more interesting from the point of view of biological applications. It can be used to describe various compositional aspects of the population.

(c) $\qquad$ For nice random characteristics χ' and χ'',

$\qquad Z_t^{\chi'} / Z_t^{\chi''} \longrightarrow E[\hat\chi'(\alpha)] / E[\hat\chi''(\alpha)]$

$\qquad$ in different senses on the set of non−extinction.

When $Z_t^{\chi''} = y_t$, i.e. the total population process, the limit in (c) reduces to $E[\hat\chi'(\alpha)]$. The importance of this result comes from the fact that if χ'' is the indicator function of some interesting property, and χ' associates with every individual having this proprty a (life− and age dependent) contribution, then the above empirical ratio can be written in the form of an empirical mean

$$(1 / Z_t^{\chi''}) \sum_{x\in I} \chi'_x(t-\sigma_x).$$

As in elementary statistics, such an empirical mean can can be interpreted as the empirical expectation of the contribution described by χ' over the group of individuals having the property specified by χ'' at time t. Equivalently we can say that if we sample one individual at random among all those who have the property specified by χ'' (i.e. if every individual in this group is chosen with a probability equal to one divided by the number, $Z_t^{\chi''}$, of the individuals in the group), then the ratio above can be understood as the expected contribution, at time t, of this randomly sampled individual. Since all individuals having the property in question are sampled with equal probabilities, we can view the random individual as typical for this subpopulation. An elementary fact from probability theory says that the expectation of the indicator of an event equals the probability of this event. Applied to the above ratio, this says that if χ' is the indicator function of the property described by χ'' and some other property , then the ratio can be interpreted as the probability that the random individual possesses this second property. Another formulation of the same idea is to say that the above ratio describes the proportion of individuals having the second property among those having the first one. Using (c) we are also able to describe the corresponding proportion in a very old population or equivalently the asymptotic probability related to a randomly sampled individual. If we fix χ'' and vary χ', we obtain various probabilities related to a randomly sampled individual chosen among those in the subcategory described by χ'' (for more precise formulations cf. Chapter Five). Let us consider an explicit example.

EX. 3   If χ'' is as in EX. 2 and
$$\chi_x'(t-\sigma_x) = \begin{cases} 1 \text{ if } a \leq t-\sigma_x \leq \lambda_x \\ 0 \text{ otherwise} \end{cases}$$
then the ratio $Z_t^{\chi'} / Z_t^{\chi''}$ will give the proportion of individuals older than age a among those alive at time t or equivalently the probability that the age of a (living) individual sampled at random at time t is larger than a. According to (c) this probability will converge to the so called stable age distribution $\dfrac{E[\hat{\chi}'(\alpha)]}{E[\hat{\chi}''(\alpha)]}$ where

$$E[\hat{\chi}'(\alpha)] = \int_0^a \alpha e^{-\alpha u} P(\lambda > u) du,$$

$$E[\hat{\chi}''(\alpha)] = \int_0^\infty \alpha e^{-\alpha u} P(\lambda > u) du,$$

and λ stands for the life span.

Fortunately, the above intuitive reasoning concerning individual random characteristics extends to more general ones. We have already stated that a random characteristic may depend on the life histories of the progeny of a counted individuals ,but we can in fact allow them to depend on the life,histories of her other relatives (ancestors, cousins, uncles etc). If χ' is of the latter type, we can use results similar to (c) to write down any asymptotic probability distribution related to the (forward– and backward–) pedigree of a random individual sampled from some interesting subcategory of individuals.

The most basic fact about supercritical branching processes is that they either become extinct or explode at an exponential rate. As explained above, it is still possible to imagine some kind of asymptotic stability (given non–extinction), namely that of the composition, in various senses, of the population according to some appropriate limiting composition laws even if the population size continues to grow. We have also seen that an alternative way of thinking about such composition laws is to view them as related to some randomly sampled individual chosen from some subcategory of individuals in a very old branching population. As we will explain in Chapter Five, it is possible to summarise the above limit distributions into one single probability measure which we call the stable pedigree law, and which is induced by the life law and the sampling mechanism. In the sequel, the abbreviation RSI will be used to denote a randomly sampled individual chosen among all those born in a very old branching population, both dead and alive. The stable pedigree measure resulting from sampling in this maximal category will be denoted by $\tilde{P}$. In Chapter Five, we show how the stable pedigree measure valid for any other sampling scheme based on some individual property can be obtained from $\tilde{P}$. In the important special case when we sample among those still alive at the sampling time point, the corresponding measure will be denoted by $\tilde{P}_\ell$. According to the above discussion, there are two equivalent ways of obtaining limit results when dealing with branching processes: one via ratios of branching processes counted by random characteristics, and the other by a direct use of the stable pedigree measure.

0.2. THE NEUTRAL THEORY

Let us now say a few words about our second ingredient, the neutral theory of molecular evolution.

The advent of molecular biology has led to new insights into the nature of the process of evolution of living organisms. It became for example possible to study this process at its most basic level, that of nucleic acid– and amino acid sequences, instead of studying it at the level of the phenotypic consequences of those. In this light it became meaningful to compare distant species. The famed molecular clock hypothesis relies on the belief that point mutations at some specific site of the DNA accumulate at an approximately steady rate over time in most distant species. This seems to suggest a clocklike relationship between numbers of mutations and absolute time. In reality, however, this idealized relationship is not always compatible with empirical data. The tick–tack of the molecular clock (if there is any molecular clock at all) seems to be

When confronted with such facts, some scientists were led to the provocative idea of neutral mutations i.e. inheritable changes that are neither beneficial nor harmful for the organism (cf. [Kimura, 1968] and [King & Jukes, 1969]). This idea contained the germ of what is now known as the neutral theory of molecular evolution.

The main feature of the neutral theory is the assertion that a mutant allele, which functions in about the same way as the modified one, may very well become established in the population due to sampling effects at reproduction. Molecular evolution can thus, according to the neutral theory, partly be accounted for in terms of accumulation of neutral mutations.

The impact of this new theory on population genetics is obvious. There is a large body of research papers where mathematical models are suggested in order to account for the behaviour of "neutral populations" (cf. [Tavaré, 1984] and [Ewens, 1990] for references). Among other things, this has led to new interesting problems not only for the population geneticist but also for the mathematician and the probabilist. Most of those mathematical models are based upon the idea that since the number of nucleotide sites in a gene is large, the number of possible alleles is, at the molecular level, practically infinite. Therefore it is not unrealistic to assume that every mutation will lead to a new allelic form never encountered in the population before. Such models are called infinite alleles models and were initially considered by [Kimura & Crow, 1964]. A slightly different kind of model is obtained if we consider point mutations affecting only one nucleotide site of the gene. Since the number of sites is large, every new mutation will almost always affect a new site never corrupted before. Such models are called infinite sites models and were introduced by [Kimura, 1969].

0.3.   THE RESULTS

From now on, we consider a branching process of the above described type to which we add mutation indices ρ: N×Ω → {0,1} describing the genetic status of the offspring as follows

$$\rho(k, \omega_x) = \begin{cases} 0 & \text{if x's kth child is a (neutral)} \\ & \text{mutant (in which case she carries} \\ & \text{an entirely new label)} \\ 1 & \text{if x's kth child carries the same} \\ & \text{label as x} \end{cases}$$

We call the resulting process a labelled branching process. Notice that the probability of the event "x is mutant" is a flexible function of the life of x's mother and may for example vary with the age of this mother. Obviously, this allows for an arbitrary dependence between the mutation indices of siblings. Taking the mutation probability to be some constant θ between zero and one independently of everything else provides the common assumption of the population genetics literature.

It is intuitively clear that we can split the reproduction point process, ξ, into ξ_m and ξ_n describing the individual reproduction of mutant and non–mutant offspring respectively (cf. Chapter Two). Obviously

$$\xi(t) = \xi_m(t) + \xi_n(t) \quad \text{and} \quad \mu(t) = \mu_m(t) + \mu_n(t),$$

where $\mu_m(t) = E[\xi_m(t)]$ and $\mu_n(t) = E[\xi_n(t)]$.

The main body of the book begins with a first chapter devoted to a detailed (technical) description of general branching processes.

In Chapter Two we give a formal definition of the labelled branching process and discuss the asymptotics of the branching process obtained when ξ is replaced by ξ_n. This new process (to be called the ancestral process) removes the lines of descent initiated by mutants. In other words, the ancestral process contains all individuals carrying the same label as the ancestor (the ancestral label) and nobody else. It can thus be used to study the fate of this ancestral label. It is not difficult to realize that the fate of the ancestral label will say something about the destinies of other labels.

Apart from the original problem of the persistence of family names, the very first use of branching processes is due to Fisher (cf. [Fisher, 1922]). In that reference, Fisher tries to answer the question: Under what circumstances will the progeny of a new mutant gene become extinct ? The same question makes sense in the context of our model. The answer is the following. The ancestral label will (with probability one) disappear from the population if $\mu_n(\infty) \leq 1$ while it will survive with a positive probability whenever $\mu_n(\infty) > 1$. In the case where $\mu_n(\infty) \leq 1$, it is also possible to describe the tail of the distribution of the time, T, it takes the ancestral label to become extinct:
- Under suitable conditions and provided $\mu_n(\infty) < 1$,

$$P(T>t) \sim c_1 e^{\alpha_n t}, \quad \text{as } t \to \infty$$

where c_1 is some positive constant and α_n solves the equation $\hat{\mu}_n(\alpha_n) = 1$.
- If on the other hand $\mu_n(\infty) = 1$, we have the tail probability

$$P(T>t) \sim c_2/t, \quad \text{as } t \to \infty$$

where $c_2 = 2 \int_0^\infty u\, \mu_n(du)/\text{Var}[\xi_n(\infty)]$ under some natural assumptions.

In Chapter Three, we are concerned with the number, N_t, of labels ever seen in the population up to some time point t and the number, M_t, of labels still represented by at least one living individual at time t. Since both N_t and M_t can be written as branching processes counted by random characteristics, we can make use of some basic convergence theorems (cf. APPENDIX A) for such processes to get the approximations

$$E[N_t] \sim e^{\alpha t} \hat{\mu}_m(\alpha)/\alpha\beta$$

$$\mathrm{Var}[N_t] \sim \frac{e^{2\alpha t} \hat{\mu}_m(\alpha)^2 \mathrm{Var}[\hat{\xi}(\alpha)]}{\alpha^2 \beta^2 (1-\hat{\mu}(2\alpha))}$$

as $t \to \infty$. Under condition (6) we can also conclude the convergence

$$N_t/y_t \to \hat{\mu}_m(\alpha) \quad \text{with probability one on the set} \quad \{y_t \to \infty\}.$$

Similar results are obtained for $E[M_t]$ and M_t/y_t. An analogous of $\mathrm{Var}[N_t]$ for M_t is given in Chapter Four. Chapter Three finishes with an application and a discussion of the lattice case. When the underlying branching process is of the Galton–Watson type (individuals reproduce at age one which is also their death moment), we obtain some of the results of [Pakes, 1984].

For many purposes it is desirable to be able to deal with the labels themselves viewed as the (generalized) individuals of some branching population which is embedded in the underlying one. We call this process the label process. In Chapter Ten we present a device for regrouping individuals in a branching population in such a manner that the resulting groups behave as (generalized) individuals in a branching process which is embedded in the original one. In this light the label process is the embedded process where the groups are the different subsets of I containing individuals with the same label. Technically, this is done using the concept of branching processes defined on spaces of marked trees as defined in [Neveu, 1986]. Chapter Ten is very technical and may be omitted by readers who are not very interested in mathematics. We use this embedding idea in the first section of Chapter Four to characterize the label process and check its counterparts of the requirements (1) − (6) above. In the two remaining sections of that chapter we give two applications.

In the first application, we show that the point process of the ages of the oldest labels, when translated in an appropriate manner, becomes mixed Poisson in character as time increases (in the sense of weak convergence of point processes).

The second application is related to the question: How frequent are the most frequent alleles?

Both applications use a distributional limit theorem proved in [Jagers & Nerman, 1984b] for branching processes counted by random characteristics which may depend on absolute time (as opposed to random characteristics which may only depend on life histories and ages).

So far we have mainly discussed prospective aspects of the labelled branching process. Besides Chapter Ten, the remaining chapters of the book are devoted to retrospective aspects. More precisely, we describe what happens when we enter a very old branching process and use the stable pedigree measures, $\tilde{P}$ and $\tilde{P}_\ell$, to study some time reversal aspects of the pedigree of a randomly sampled individual.

In Chapter Five some new notation is given as well as (formal) descriptions of the stable pedigree measure, $\tilde{P}$, and the stable pedigree space on which $\tilde{P}$ acts. We also explain how $\tilde{P}_\ell$ relates to $\tilde{P}$. The whole chapter is more or less based on [Nerman & Jagers, 1984].

Chapter Six begins by the study of the generation age, ν, of the label of the RSI (cf. §0.1 for the convention defining the abbreviation RSI). All the results of Chapter Six are also true under $\tilde{P}_\ell$, i.e. when we sample among those alive. ν is defined as the number of backward generations since this label arose by mutation. ν turns out to be geometrically distributed with parameter $\hat{\mu}_m(\alpha)$, this being the $\tilde{P}$ − probability that the RSI (or any of her ancestors) is a mutant. Similary we study the real time age, T, of the label of an RSI, defined as the length of the backward time interval between the birth moment of the RSI and that of her last mutant ancestor. We are able to show that

With $\mu_\alpha^n(t) = \int_0^t e^{-\alpha u} \mu_n(du)$, we have

(a) $\qquad \tilde{P}(T \leq t; \nu = k) = (\mu_\alpha^n)^{*k}(t) \hat{\mu}_m(\alpha)$

(b) $\qquad \tilde{P}(T \leq t) = \sum_{k=0}^{\infty} (\mu_\alpha^n)^{*k}(t) \hat{\mu}_m(\alpha)$

(c) $\qquad \tilde{E}[T] = B_n / \hat{\mu}_m(\alpha)$

where $B_n = \int_0^\infty u e^{-\alpha u} \mu_n(du)$ and $\tilde{E}$ denotes expectation with respect to $\tilde{P}$.

The remaining part of Chapter Six is devoted to the study of the backward process of the birth moments of the mutant ancestors of the RSI. We let $X(s)$ be the number of mutant ancestors of the RSI during the interval $[-a, -a-s]$ where we think of the sampling time point as the time origin and a as the age of the RSI at the sampling moment.

It turns out that $X(s)$ behaves as a delayed renewal process with interarrival time distribution function

$$F(t) = \sum_{k=0}^{\infty} \mu_\alpha^m * (\mu_\alpha^n)^{*k}(t)$$

and delay time distribution function

$$F_T(t) = \sum_{k=0}^{\infty} (\mu_\alpha^m)^{*k}(t) \hat{\mu}_m(\alpha).$$

A renewal process can be seen as a process which counts the number of customers arriving to some service station when the (random) times between the successive arrivals are independent and identically distributed. The word delayed means in this context that the time until the first arrival is stochastically different from the remaining interarrival times. In our case, the interarrival times coincide with the lengths of the time intervals between the birth times of the successive ancestors. Standard results about such renewal processes can be used to describe the asymptotics of X(s) and other related processes. This is done in THEOREM (6.5).

In Chapter Seven the purpose is to define a measure of variation over labels within the population when this has existed long enough. The usual measure of relatedness used in the population genetics literature is the probability of identity by descent or equivalently the probability that two randomly sampled individuals carry the same label. We define such a measure by first considering the mean size, φ, of the label of an RSI, the size of a label being the number of individuals carrying that particular label. This is done in THEOREM (7.1), the main idea being:

Let φ be as above and assume that requirement (6) holds. Then if $y_n(t)$ denotes the total of number individuals born before t (both dead and alive) in the ancestral process, then

$$\tilde{E}[\varphi] = \hat{\mu}_m(\alpha) \int_0^\infty \alpha e^{-\alpha t} E[y_n(t)^2] dt < \infty,$$

if we can find a γ meeting the following requirements:

(i) $\qquad\qquad \gamma < \frac{\alpha}{2}$

(ii) $\qquad\qquad E[\hat{\xi}_n(\gamma)] < 1$

(iii) $\qquad\qquad E[\hat{\xi}_n(\gamma)^2] < \infty$

The form of $\tilde{E}[\varphi]$ agrees with the main result of §7.2:

$$\tilde{P}(\varphi=j) = \hat{\mu}_m(\alpha) \int_0^\infty \alpha e^{-\alpha t} j P(y_n(t) = j) dt.$$

The corresponding result in the case where we sample among those alive is

$$\tilde{E}_\ell[\varphi] = \hat{\mu}_m(\alpha) \int_0^\infty \alpha e^{-\alpha t} E[(Z_t^n)^2] dt / (1 - \hat{L}(\alpha)),$$

where $L(u) = P(\lambda < u)$, λ being the life span of an individual and Z_t^n stands for the number of living individuals at time t in the ancestral process (see above).

The main finding of Chapter Seven is the following. Let P_t denote the conditional probability of the event that two randomly sampled individuals chosen with replacement among those alive at time t are identical by descent, given the branching population. Let Z_t stand for the number of living individuals at time t. Then P_t can be approximated by the population size determined expression

$$P_t \sim \tilde{E}_\ell[\varphi]/Z_t.$$

In the infinite sites model, every label is identified with an infinite sequence of sites and every new mutation acts on a new site never mutated before. A first question arises. What is the mean number of individuals differing from a randomly sampled individual with respect to exactly d sites? This and other related questions are considered in Chapter Eight.

In Chapter Nine we answer the questions:

(i)   What information about the label of an individual, x, do we gain from some knowledge about the label of another individual, y, related to x in a specific manner?

(ii)   How is x likely to be related to y given some knowledge about the genetic (label) constitution of x and y or at least their relative genetic relationship?

Chapter Eleven is more speculative and is devoted to various generalizations and extensions as well as some suggestions for further research.

In what follows, we mention some references containing branching process models similar to the one studied in this book. [Pakes, 1984] and [Pakes, 1987] treat the Galton-Watson and the so called Markovian cases respectively, using a different approach. Of course, such models are special cases of our general labelled branching process model, but it should be mentioned that the emphasis in Pakes' papers is on other aspects of labelled branching processes than those we are interested in. [Tavaré, 1989] (cf. also [Karlin & McGregor, 1967]) proposes a birth-and-death population model subject to a stream of (neutral) mutants joining the population according to a time homogeneous Poisson process. Finally we mention [Sawyer, 1976] where a branching process with a geographical structure is discussed and the two papers by Bühler ([Bühler, 1971] and [Bühler, 1972]) formulated in terms of general genealogical problems.

CHAPTER ONE

THE CONSTRUCTION OF THE PROCESS

The modern construction of the general branching process relies heavily upon the Ulam-Harris family history space. The following version is mainly based on [Jagers & Nerman, 1984a] (for another construction cf. Chapter Ten). The idea of using random characteristics was introduced in [Jagers, 1974] and was later on developed in [Nerman, 1981] and [Jagers & Nerman, 1984a] to become a useful tool in the study of various aspects of branching populations counted in different manners.

1.1. THE BASIC PROBABILITY SPACE

Let us agree to call every vector

$$x = (x_1,..., x_n) \in N^n$$

$$N = \{1,2,...\}$$

$$n \in N$$

an individual in the nth generation. The interpretation of this is that x is the x_nth child of the x_{n-1}th child... of some founder individual, 0, to be called the ancestor of the population. We shall assume that the population of individuals in all generations is initiated by this single ancestor. The countable set $I = \bigcup_{n=0}^{\infty} N^n$, where $N^0 = \{0\}$, the Ulam-Harris individual space, is easily seen to be large enough to "label" whatever empirical populations we may study. If x and y are individuals in the nth and the mth generations respectively, then we shall agree to write xy for the (n+m)th dimensional vector (= individual) whose n first components are those of x and whose m last components are those of y. This, together with the convention that $x0 = 0x = x$, defines a composition law for individuals.

For most of our genetical applications, it is quite satisfactory to describe the different aspects of the life history of an individual by some vector of random entities (λ, τ, ρ) where λ stands for the life span of the actual individual, τ for the different ages at childbearing and ρ describes whether her children are mutants or not. However, to enjoy greater freedom in dealing with the concept life history (or life for short as we will call it in the sequel), we adopt a more general point of view and assume that we are given a measurable space $(\Omega, \mathscr{A})$, the life space, rich enough to allow us to define on it (at least) the above mentioned entities, i.e.

-    The life span $\lambda: \Omega \to R^+ \cup \{+\infty\}$.

-    The different ages at childbearing, given by

$$\tau: N \times \Omega \to R^+ \cup (+\infty)$$

with the interpretation that $\tau(k,\omega)$ is the age at the birth moment of her kth child of a mother whose life is described by $\omega \in \Omega$. $\tau(k,\omega)$ is taken to be infinite if the mother in question never begets k children. The point process (cf. [Kallenberg, 1983] for a general approach or the Appendix in [Leadbetter et al, 1983] for point processes on the line) defined by $\xi(A) = \#\{k \in N; \tau(k,\omega) \in A\}$, where A is any real Borel set, is to be named the reproduction process. This process provides information about the different ages at childbearing and thereby also information about the numbers of offspring born in different "age sets". The above definition of ξ does not exclude the possibility of allowing dead individuals to reproduce but for us this will only be permitted in the sense that $\lambda(\omega) < \tau(k,\omega)$ implies that $\tau(k,\omega) = +\infty$. In Chapter Ten we account for another (minimal) construction which excludes such non-realized offspring.

- The mutation indices

$$\rho: N \times \Omega \to \{0, 1\}$$

are to be interpreted according to the following: $\rho(k,\omega)$ takes the value zero if the kth child of a mother whose life is described by $\omega \in \Omega$ is a mutant (i.e. differs from her in some selectively neutral manner that does not affect her ability to survive and reproduce) and the value one otherwise.

All the above mappings are of course assumed to be measurable.

We now equip the life space with some probability measure P, the life law, to get our basic outcome space

$$(\Omega^I, \mathscr{A}^I, p^I) = \prod_{x \in I} (\Omega, \mathscr{A}, P)$$

the population process. Doing so, we make a crucial assumption about the lives of distinct individuals being i.i.d. (independent and identically distributed). For ease of notation, P^I will be replaced by P and we will let ω denote an outcome of the process.

The evolution of the population in time can be caught through the birth time points σ_x recursively defined by

$$\sigma_0 = 0 \quad \text{(the ancestor is born at the time origin)}$$

and

$$\sigma_{xk} = \sigma_x + \tau(k, \omega_k)$$

where ω_x denotes the coordinate projection in Ω^I (i.e. if p_x is the projection map of Ω^I onto Ω we will have $p_x(\omega) = \omega_x$ where ω denotes an outcome of the population process). An important role will be played by the class of shift (or projection) operators

$$S_x: \Omega^I \to \Omega^I, \quad x \in I$$

mapping $\{\omega_y : y \in I\}$ to $\{\omega_{xy} : y \in I\}$. The name the daughter process of x given in [Jagers & Nerman, 1984a] to S_x reflects the fact that S_x "transforms" x into an ancestor. If f is any function on Ω^I, then by f_x we will mean the function on Ω^I defined by $f_x = f \circ S_x$, reservation made for some special functions where the subscript x has some otherwise defined meaning. It then follows that $f \circ S_0 = f_0 = f$. We will also have use for the following mappings on I: $n(x) =$ the dimension of the vector $x =$ the generation number of the individual x and $x_{[k]} = (x_1, ..., x_{n(x)-k})$, for $k \leq n(x)$ ($x_{[n(x)]}$ interpreted as 0). Finally we define ξ's intensity, in the point process terminology, by $\mu(A) = E[\xi(A)]$, A being a real Borel set, and adopt the comfortable notations $\xi(t) := \xi([0,t])$ and $\mu(t) := E[\xi(t)]$.

1.2. PROCESSES COUNTED WITH RANDOM CHARACTERISTICS

A random characteristic is a real valued process $\{\chi(a)\}_{a \in R}$ defined on $(\Omega^I, \mathscr{A}^I, P^I)$ and assumed to be $\mathscr{B} \times \mathscr{A}^I$ - measurable, $\mathscr{B}$ being the real Borel algebra. For our future purposes, it will be enough to consider non-negative "characteristics" vanishing on the negative real half line (an exception to this rule is however §4.2–4.3). The characteristic χ_x associated with the individual $x \in I$ is defined by $\chi_x(a) = \chi \circ S_x(a)$, $a \in R$ and can be interpreted as the score or the contribution, at age a, of the individual x to the population count. The branching process counted by the random characteristic χ can now be defined as the sum

$$Z_t^\chi = \sum_{x \in I} \chi_x(t-\sigma_x), \quad t \in R$$

of the contributions of different individuals at their actual ages at some time point t. Some illustrative examples of such processes are

Ex. 1.   When $\chi(a) = 1_{R^+}(a)$, we get the important process $y_t = Z_t^\chi =$

$\sum_{x \in I} 1_{R^+}(t-\sigma_x)$ counting all individuals born before t. This process is given the name the total population process.

Ex. 2.   When $\chi(a) = 1_{[0,\lambda_0]}(a)$, we get the process Z_t counting indivuduals

alive at time t.

Ex. 3.   The characteristic $\chi(a) = 1_{[0,\lambda_0 \wedge b]}(a)$ gives the process Z_t^b counting

individuals alive and younger than b at t (where u ∧ v means the minimum of u and v).

Ex. 4.   Taking $\chi(a) = 1_{R^+}(a) 1_N(\xi(a))$ yields the process counting mothers with at least one child born before or at their actual ages.

A random characteristic is said to be individual or of the individual type if, as a function on Ω^I, it does not involve the lives of other individuals than the ancestor. Then for any $x \in I$, χ_x can be written as a function of ω_x alone. The characteristics involved in the above examples were all of the individual type. We give a final example of a characteristic which is non-individual.

Ex. 5.   Taking
$$\chi(a) = 1_{R^+}(a - \max_{i \in N} \{\tau(i, \omega_0) + \lambda(\omega_i)\})$$
yields the process Z_t^χ counting the number of mothers with all their children dead at age a.

1.3.   SPECIAL CASES

Sometimes it is preferable to deviate from the general case and consider special classes of branching processes. It turns out that this is easily done by specializing the reproduction process and the life span. We content ourselves by giving two examples.

A. THE BELLMAN-HARRIS PROCESS.

Let the life span λ be distributed according to some distribution function G_λ. Assume further that every individual begets, at death, a (random) number, B, of children and define $p_k = P(B=k)$, $k \in N \cup \{0\}$. With λ and B taken to be independent, and
$$\xi(a) = \begin{cases} B & \text{if } a \geq \lambda \\ 0 & \text{if } a < \lambda \end{cases}$$
we get $\mu(a) = E[\xi(a)] = b\, G_\lambda(a)$ where $b = \sum_{k=0}^{\infty} p_k k$. The pairs (λ_x, ξ_x), $x \in I$ are assumed to be the lives, but we can without any problem use our richer life concept. The resulting process is usually called the Bellman-Harris branching process.

B. THE GALTON-WATSON PROCESS

This process is obtained by taking $\lambda(\omega) = 1$ for all $\omega \in \Omega$ and assuming that an individual splits at age 1 into a (random) number, B, of offspring. If again $b = E[B]$, we can write
$$\xi(a) = \begin{cases} B & \text{if } a \geq 1 \\ 0 & \text{otherwise} \end{cases}$$
and
$$\mu(a) = \begin{cases} b & \text{if } a \geq 1 \\ 0 & \text{otherwise.} \end{cases}$$

1.4. DECOMPOSITIONS

As already mentioned, a crucial assumption on branching populations is that the lives of distinct individuals are assumed to be i.i.d. This is often referred to as the single-type branching property. In [Jagers & Nerman, 1984a] three basic lemmas are stated as immediate (!) consequences of the branching property and are used in the treatment of some useful time- and generation-wise decompositions of the population counts (formal proofs of these lemmas are to be found in [Chauvin, 1986]). If we define the daughter process of x counted by χ as $Z_t^\chi(x) = Z_t^\chi \circ S_x$ and take $I(t) = \{xk;\ \sigma_x \leq t < \sigma_{xk} < \infty,\ t \in R^+\}$ (in words $I(t)$ is the set of still not born offspring pertaining to already born mothers), we get

$$Z_{t+u}^\chi = \sum_{\substack{x \in I \\ \sigma_x \leq u}} \chi_x(t+u-\sigma_x) + \sum_{x \in I(u)} Z_{t+u-\sigma_x}^\chi(x)$$

which can be called the time-wise decomposition and

$$Z_t^\chi = \sum_{k=0}^{n-1} \sum_{x \in N^k} \chi_x(t-\sigma_x) + \sum_{x \in N^n} Z_{t-\sigma_x}^\chi(x)$$

to be called the generation-wise decomposition. An important special case of the latter decomposition is the basic decomposition according to ancestry in the first offspring generation

$$Z_t^\chi = \chi_0(t) + \sum_{i \in N} Z_{t-\tau(i,\omega_0)}^\chi(i).$$

For more details the reader is referred to [Jagers & Nerman, 1984a]. Such decompositions will be made use of in the sequel without further justifications.

1.5. GENERAL ASSUMPTIONS

The branching process defined in the previous section is often referred to as the general branching process (also the name Crump-Mode-Jagers (C-M-J) is sometimes used). The attribute general is justified by the fact that no special patterns of individual behaviour are assumed and of course no explicit distributions. This freedom is however to be restricted if we are to be able to "work" with the process. Throughout the present section we state a number of rather technical assumptions and conditions ensuring the good behaviour of the process and the applicability of the basic convergence theorems stated in APPENDIX A. Before doing so we have to underline the well known fact that branching processes either die out or continue to grow forever. In our case we will mainly consider such processes that may escape extinction with positive probability. This assumption is formally stated in condition (C.1) below. Branching processes satisfying (C.1) are called supercritical.

(C.1) $\qquad \mu(\infty) = E[\xi(\infty)] > 1$ (supercriticality).

(C.2) $\exists \alpha \in R$ such that $\hat{\mu}(\alpha) = 1$. Such an α is called the Malthusian parameter. This condition means that the reproduction measure can be "normed" to a probability measure, and implies among other things the applicability of the key renewal theorem. The notation $\hat{f}(\alpha)$ will be retained to denote the Laplace-Stieltjes transform of f generally. (i.e.

$$\hat{f}(\alpha) = \int_0^\infty \alpha e^{-\alpha t} f(t) dt).$$

(C.3) $\beta = \int_0^\infty u e^{-\alpha u} \mu(du) < \infty$. Processes satisfying this condition and (C.1) are

called Malthusian. β is usually given the name the mean age at childbearing.

(C.4) μ is non-lattice. This means that μ cannot be supported by any set

$$\{0, d, 2d,...\}, \quad d > 0.$$

This is a well known condition from renewal theory. Processes with reproduction measures supported by such lattices are considered in APPENDIX B and §3.4.

(C.5) $\exists \alpha' < \alpha$ such that $\hat{\mu}(\alpha') < \infty$. Notice that $\alpha' < \alpha \Rightarrow \hat{\mu}(\alpha') > \hat{\mu}(\alpha)$.

(C.6) $\mathrm{Var}[\hat{\xi}(\alpha)] < \infty$.

(C.7) $E[\hat{\xi}(\alpha)\log^+\hat{\xi}(\alpha)] < \infty$. This is usually called the "x logx" condition.

Processes satisfying (C.1)-(C.4) will be called "well behaved". The remaining assumptions will be referred to when needed.

Notice that the above assumptions are by no means completely independent from one another. However, it turns out to be convenient to list them the way we do since we are planning to use them in their actual forms. Condition C.5 is not stronger than C.6 as it may appear at first sight. In fact none of these two conditions implies the other one quite generally as can be seen by counter-examples. As usual, we will call processes satisfying $\mu(\infty) = 1$ critical and processes satisfying $\mu(\infty) < 1$ subcritical.

NOTE. This section is intimately related to APPENDIX A where the basic convergence theorems for general branching processes are stated for ease of reference. In that appendix are also stated some regularity conditions on random characteristics χ. It may thus help to have a look at those theorems already at this stage.

CHAPTER TWO

LABELLED BRANCHING PROCESSES

In this chapter we give the formal definition of a labelled branching process and start to study one of its aspects: the fate of the label carried by the ancestor. Using quite well known results we are able to describe the extinction and the growth of the subpopulation of individuals carrying this label. Some of the formulae given in this chapter will be made use of in §4.2.

2.1. LABELLED BRANCHING POPULATIONS

In Chapter One we describe a general branching process to which we add neutral mutation indices. We will agree to call such a process a labelled branching process. Think of a labelled branching population as a model for the development of a haploid population where the offspring can undergo neutral mutations according to the infinite alleles hypothesis. In other words every individual inherits an allele (or a label) which she either transmits or not to her offspring. An individual not receiving her mother's label is called a mutant and is assumed to carry an entirely new label never seen in the population before. Individuals carrying distinct labels are assumed to behave equally well, at least with respect to the vital aspects of their lives such as reproduction, viability, etc.

The mutation indices can be used to split the reproduction point process, ξ, into $\xi_m(t)$ (m for mutant) and $\xi_n(t)$ (n for non-mutant) giving the number of mutant and non–mutant children born to the ancestor during the time interval $[0,t]$. Obviously

$$\xi_n(t) = \sum_{i \in N} \rho(i,\omega_0) 1_{[0,\infty)}(t - \tau(i,\omega_0)),$$

and

$$\xi(t) = \xi_n(t) + \xi_m(t).$$

Doing so reflects the fact that we are not interested in the labels themselves but in the "variation over labels" within the population. Alternatively, ξ could have been defined as a marked point process where every occurrence point is given a mark which is either of 0 or 1 and the sequence $\{\rho(k)\}_{k \in N}$ gives the successive marks. Finally, we denote the intensities of ξ_n and ξ_m by $\mu_n(t) = E[\xi_n(t)]$ and $\mu_m(t) = E[\xi_m(t)]$. Then of course $\mu = E[\xi] = \mu_m + \mu_m$.

In what follows, we will refer to the branching process using ξ as reproduction point process as the underlying process. This underlying process will always be assumed to be well behaved in the sense of §1.5.

2.2. THE FATE OF THE ANCESTRAL LABEL

An interesting question, and one which can be answered immediately, is "what destiny is the ancestral label (i.e. the label carried by the ancestor of the population) likely to meet"? Naturally, the fate of the ancestral label will say something about the destinies of the other labels.

More exactly, we want to discuss two problems concerning the ancestral label: its possible disappearance on one hand and its prevalence on the other.

It is not hard to realize that the process of individuals carrying the ancestral label, which we call the ancestral process, is again a general branching process which can be obtained from the original one simply by replacing ξ by ξ_n.

Let $q_n(t)$ and q_n denote the probabilities of extinction of the ancestral process before t and ultimately. Then with

$$Z_t^n = \text{the number of living individuals carrying the ancestral}$$

label at time t,

we have

$$q_n(t) = p(Z_t^n = 0),$$

and it can be proved that q_n is the smallest positive root of the equation

$$s = E[s^{\xi_n(\infty)}].$$

It can also be proved that, except in trivial cases,

$$\mu_n(\infty) \leq 1 \;\Rightarrow\; q_n = 1$$

$$\mu_n(\infty) > 1 \;\Rightarrow\; q_n < 1$$

and that the form of $q_n(t)$ can, in some few cases, be obtained by solving the so called Kolmogorov's backward equation (cf. [Athreya & Ney, 1972]).

In the subcritical or ciritical cases (i.e. when $\mu_n(\infty) \leq 1$), some well known results (cf. [Asmussen & Hering, 1983]) allow us to consider the asymptotic distribution of what might be called the time, T, until extinction of the ancestral process, i.e.

$$T = \inf\{t; Z_t^n = 0\}.$$

THEOREM (2.1) (The subcritical case)

Assume that $\quad\quad$ (1) $\quad \mu_n(\infty) < 1$

$\quad\quad\quad\quad\quad\quad\quad$ (2) $\quad \exists \alpha_n$ such that $\dot{\mu}_n(\alpha_n) = 1$

$\quad\quad\quad\quad\quad\quad\quad\quad\quad$ (such an α_n satisfies $\alpha_n < 0$).

$$(3) \quad \hat{G}_\lambda(\alpha_n) < \infty \quad (G_\lambda(t) = P(\lambda \leq t))$$

$$(4) \quad \int_0^\infty u e^{-\alpha_n u} \mu_n(du) < \infty$$

$$(5) \quad \mu_n \text{ is non-lattice.}$$

Then

$$P(T > t) \sim c e^{\alpha_n t}$$

where c is under "x log x" for the ancestral process a positive constant which depends on the distribution of (ξ_n, λ) in a complicated manner. If "x log x" fails then $c = 0$.

THEOREM (2.2) (The critical case)

Assume that

$$(1) \quad \mu_n(\infty) = 1 \text{ (in this case } \alpha_n = 0)$$

$$(2) \quad \lim_{t \to \infty} t^2(1 - \mu_n(t)) = 0$$

$$(3) \quad \lim_{t \to \infty} t^2(1 - G_\lambda(t)) = 0$$

Then with $\sigma_n^2 = \text{Var}[\xi_n(\infty)]$ and $\beta_n = \int_0^\infty u \mu_n(du)$ we get

$$P(T > t) \sim 2\beta_n / \sigma_n^2 t,$$

assuming β_n and σ_n^2 to be finite.

The two above theorems will come to use in Chapter Four. The growth of the ancestral process can be described by the usual convergence theorems available in the literature. In order to illustrate this we notice that under suitable conditions:

- In the subcritical case

$$\lim_{t \to \infty} P(Z_t^n \leq u | Z_t^n > 0) = D(u),$$

in the sense of weak convergence of distribution functions, where D is a proper distribution function which satisfies

$$\int_0^\infty t D(dt) = (1 - \hat{G}_\lambda(\alpha_n))/\beta_n c$$

and c is as in THEOREM (2.1).

- In the critical case

$$\lim_{t \to \infty} P(Z_t^n / t \leq u | Z_t^n > 0) = 1 - \exp[-2\beta_n u / E[\lambda] \sigma_n^2].$$

CHAPTER THREE

THE NUMBER OF DISTINCT LABELS

Let us now consider the number of labels encountered in the population in several senses. Thanks to the idea of counting subpopulations of a general branching process by means of random characteristics, we are able to study not only the total number of distinct labels ever seen in the process up to some time point   t,   but also the development of the cardinalities of label groups exhibiting some required properties. As a natural example of such counts, we shall consider the set of labels represented by at least one living individual at   t.   This will be referred to as the set of "present labels" at   t.

Let N_t stand for the total number of labels and M_t for the number of present labels at time   t.   In the Galton-Watson case [Pakes, 1984] contains results about the convergence of $b^{-n}E[M_n]$ and the a.s. convergence of M_n/Z_n where Z_n and M_n stand for the size of and the number of different labels in the nth generation respectively. As usual,   b   stands for the mean reproduction of an individual. Similar results will be derived in this chapter in the context of our more general model.

3.1.   THE TOTAL NUMBER OF LABELS

Define N_t as the total number of labels ever encountered in the process up to time   t,   except the ancestral label. The key idea in the study of N_t is that this process can be written as a branching process counted by a random characteristic, i.e.

$$N_t = Z_t^\chi = \sum_{x \in I} \chi_x(t-\sigma_x), \quad t \in R^+$$

for χ as defined in the proof of the following proposition where   *   stands for convolution.

PROPOSITION (3.1)

Define $\nu(t) = E[y_t]$, y_t being the total population process (cf. Chapter One). In finite time N_t satisfies $E[N_t] = \mu_m * \nu(t), \quad t \in R^+$.

PROOF

The following is essentially the proof of Theorem 3.1 in [Jagers & Nerman, 1984a]. Consider the function:

$$\chi'_x(u) = \begin{cases} 1 & \text{if } x \text{ is mutant and } u \geq 0 \\ 0 & \text{otherwise} \end{cases}.$$

Clearly, with $n(x) = n$ when $x \in N^n$ and $x_{[k]} = (x_1, ..., x_{n(x)-k})$ for $k \leq n(x)$, $\chi'_x(u)$

can be written as

$$\chi_x'(u) = (1 - \rho(x_{n(x)}, \omega_{x_{[1]}}))\, 1_{R^+}(u),$$

and

$$N_t = \sum_{x \in I \setminus \{0\}} \chi_x'(t - \sigma_x).$$

However it is not possible to write

$$\chi_x' = \chi' \circ S_x$$

since χ_x' makes reference to coordinates not in S_x. Indeed the event

$$\{\rho(x_{n(x)}, \omega_{x_{[1]}}) = 0\}$$

depends on the life of x's mother. But

$$\sum_{x \in I \setminus \{0\}} \chi_x'(t - \sigma_x) = \sum_{x \in I} \sum_{i \in N} \chi_{xi}'(t - \sigma_x)$$

$$= \sum_{x \in I} \sum_{i \in N} (1 - \rho(i, \omega_x)) 1_{R^+}(t - \sigma_x)$$

$$= \sum_{x \in I} \xi_{m,x}(t - \sigma_x)$$

so that $\chi(t) = \xi_{m,0}(t)$ fits the desired scheme $(\chi_x = \chi \circ S_x)$. This resummation trick will be made use of several times in the sequel.

Define $g(t) = E[\chi(t)]$ and let $\xi^{(k)}$ be the point process that counts the birthtimes of individuals x whenever $x \in N^k$. Let further $\mathcal{A}_n$ stand for the σ-algebra generated by the lives of all individuals in $\bigcup_{0 \le k \le n} N^k$. Then, by use of a quite general method

$$E[N_t] = \sum_{k=0}^{\infty} \sum_{x \in N^k} E[\chi_x(t - \sigma_x)]$$

$$= \sum_{k=0}^{\infty} \sum_{x \in N^k} E[E[\chi_x(t - \sigma_x) \mid \mathcal{A}_k]]$$

$$= \sum_{k=0}^{\infty} E[\sum_{x \in N^k} g(t - \sigma_x)]$$

$$= \sum_{k=0}^{\infty} E[\int_0^{\infty} g(t - u) \xi^{(k)}(du)]$$

$$= E[\chi] * \sum_{k=0}^{\infty} \mu^{*k}(t),$$

where $E[\xi^{(k)}(t)] = \mu^{*k}(t)$ is easily checked by induction. The proof is now achieved by $E[\chi(t)] = E[\xi_m(t)] = \mu_m(t)$. □

It is to be noted that under (C.2) (stating that there exists an α such that $\hat{\mu}(\alpha) = 1$) $E[N_t]$ will certainly be finite and that it indeed suffices that $\mu(0) < 1$ and $\mu(t) < \infty$ to ensure the finiteness of $\nu(t)$. Under the same restriction $\text{Var}[N_t]$ is well defined. We can now let $t \to \infty$. What happens then is summarized in the following theorem.

THEOREM (3.2)

(i)   For large values of t
$$E[N_t] \sim e^{\alpha t} \hat{\mu}_{\mathbf{m}}(\alpha)/\alpha\beta$$

(ii)   If (C.6) holds, then for large values of t,
$$\text{Var}[N_t] \sim \frac{e^{2\alpha t} \hat{\mu}_{\mathbf{m}}(\alpha) \text{Var}\,[\hat{\xi}(\alpha)]}{\alpha^2 \beta^2 (1 - \hat{\mu}(2\alpha))}$$

(iii)   If (C.5) holds, then as $t \to \infty$
$$\frac{N_t}{y_t} \to \hat{\mu}_{\mathbf{m}}(\alpha) \quad \text{a.s. on the survival set } \{y_t \to \infty\}.$$

PROOF

Obviously, $E[\chi(t)] = \mu_{\mathbf{m}}(t)$ meets the requirements of theorems TH.1 - TH. 6 in APPENDIX A, and the statements follow.   □

Notice that even though the above random characteristic χ counts more than one individual, it is only a function of the life of the mother of those individuals and thereby of the individual type, which explains the applicability of TH. 2 in APPENDIX A.

3.2.   THE NUMBER OF PRESENT LABELS

Let M_t stand for the number of labels represented by at least one living individual at t with the exception of the ancestral label. Then M_t can be counted by indicators

$$\chi'_x(a) = \begin{cases} 1 & \text{if } x \text{ is a mutant with at least one living} \\ & \text{descendant with the same label as herself at age } a{\geq}0. \\ 0 & \text{otherwise,} \end{cases}$$

which again is not a true characteristic (it depends on the life of x's mother) but with the same resummation trick as in §3.1, the characteristic

$$\chi(a) =$$
$$\underset{i \in N}{\Sigma} (\,1 - \rho(i, \omega_0))1_{R^+}(a - \tau(i,\omega_0))1_{[a - \tau(i,\omega_0),\infty)}(\lambda_{\max} \circ S_i)$$

where ($\underset{k=1}{\overset{0}{\Pi}}$ interpreted as 1)

$$\lambda_{max} = \sup_{x \in I, \sigma_x + \lambda_x < \infty} \{(\sigma_x + \lambda_x) \prod_{l=1}^{n(x)} \rho(x_k, \omega_{x_1, \ldots, x_{k-1}})\},$$

yields M_t.

Let $Z_t^{\mathbf{n}}$ be defined as in the previous chapter. Then λ_{max} is the extinction time, T, of $Z_t^{\mathbf{n}}$. This can be used to prove:

THEOREM (3.3)

Recall that $q_n(t) = P(Z_t^{\mathbf{n}} = 0)$

(i) $\qquad E[M_t] \sim e^{\alpha t} \hat{\mu}_{\mathbf{m}}(\alpha)(1 - \hat{q}_n(\alpha))/\alpha\beta$ for large values of t.

(ii) $\qquad \dfrac{M_t}{y_t} \to \hat{\mu}_{\mathbf{m}}(\alpha)(1 - \hat{q}_n(\alpha))$ a.s. on $\{y_t \to \infty\}$ as $t \to \infty$, if (C.5) holds.

PROOF

Also in this case the basic convergence theorems in APPENDIX A apply immdiately. All we have to demonstrate is the equality

$$E[\hat{\chi}(\alpha)] = \hat{\mu}_{\mathbf{m}}(\alpha)(1 - \hat{q}_n(\alpha)).$$

But this is proved by

$$E[\hat{\chi}(\alpha)] = E[\int_0^\infty \alpha e^{-\alpha t} \chi(t) dt]$$

$$= \int_0^\infty \alpha e^{-\alpha t} E[\sum_{i \in N} (1 - \rho(i, \omega_0)) 1_{R^+}(t - \tau(i, \omega_0))$$

$$1_{[t - \tau(i, \omega_0), \infty)}(T \circ S_i)] dt$$

$$= \int_0^\infty \alpha e^{-\alpha t} \int_0^t P(Z_{t-u}^{\mathbf{n}} > 0) \mu_{\mathbf{m}}(du) dt$$

$$= \hat{\mu}_{\mathbf{m}}(\alpha) \int_0^\infty \alpha e^{-\alpha t} (1 - q_n(t)) dt$$

$$= \hat{\mu}_{\mathbf{m}}(\alpha)(1 - \hat{q}_n(\alpha)). \qquad \qquad \Box$$

Obviously the characteristic χ used in the above theorem is not of the individual type and consequently we cannot apply TH. 2 in APPENDIX A directly. Fortunately, this difficulty can be circumvented as we will see in Chapter Four, where the asymptotics of $Var[M_t]$ are discussed.

3.3. A BELLMAN–HARRIS APPLICATION

Assume that we are in the Bellman-Harris case (cf. §1.3). Then the Malthusian parameter is defined by $\hat{G}_\lambda(\alpha) = \frac{1}{b}$ and the supercriticality is taken care of by the condition $b > 1$, which we assume as well as the "good behaviour" of the process in the sense of §1.5. A simple way in which mutations may enter into the process is the following: During her life time a mother may (or may not) be exposed to some mutagenic factor (chemicals, radiations, etc.) in which case this happens at some (random) age η . We assume η to be distributed according to some   d.f. G_η independently of the life span. Define $\rho(i, \omega_x) = 1_{[\lambda_x \geq \eta_x]}$ and   recall that for a Bellman-Harris process $\lambda(\omega_x) = \tau(i, \omega_x)$ for all i such that $\tau(i, \omega_x) < \infty$.

In these terms a mother x with $\eta_x < \lambda_x$ can be called susceptible and her children (if she begets any) mutants. The reproduction of mutants (respectively of non-mutants) can now be described by

$$\xi_m(t) = \xi(t) \, 1_{[\eta \leq \lambda]} = B1_{[t \geq \lambda \geq \eta]}$$

and

$$\xi_n(t) = \xi(t) - \xi_m(t).$$

By taking $G(t) = G_\lambda(t)G_\eta(t) - \int_0^t G_\lambda(x)G_\eta(dx)$ we get $\mu_m(t) = b\, G(t)$. Thus $\hat{\mu}_m(\alpha) = b\, \hat{G}(\alpha)$, which can be used to formulate corollaries to THEOREM (3.2) and THEOREM (3.3).

3.4. THE LATTICE CASE

In this section we assume that time is discrete, but that the other features of the underlying process remain unchanged. If we let k denote the time index, then the population will be initiated at time $\sigma_0 = 0$ by one single ancestor 0 whose children will be born at random ages $\tau(k, \omega_0)$: $N \times \Omega \to Z^+$. In a similar manner, we can define discrete time counterparts of the variables and processes introduced so far. Also random characteristics will be defined as discrete time stochastic processes (this is of course not necessary, but it turns out to be easier to do so). There will then be no ambiguity in writing Z_k^χ , y_k , etc.

Naturally the reproduction measure $\mu(k) = E[\xi(\{0,1,...k\})]$ will now be of the lattice type and we need lattice analogs of the basic convergence theorems of APPENDIX A. Those are provided in APPENDIX B. Their proofs are mutatis mutandis rewritings of their non-lattice counterparts.

It is thus possible to formulate discrete time corollaries to the results in the preceding sections of the present chapter as well as to many of the results in the chapters to come. Before stating these, we have to notice that assumptions (C.1)-(C.4) ensuring the good behaviour of the process become

(C.1) $\quad\quad \mu(\infty) = E[\xi(\infty)] > 1 \quad\quad$ (supercriticality)

(C.2) $\quad\quad \exists \alpha$ such that $\sum\limits_{j=0}^{\infty} e^{-\alpha j}\mu(\{j\}) = 1$

(C.3) $\quad\quad \beta = \sum\limits_{j} j e^{-\alpha j}\mu(\{j\}) < \infty$ (i.e. the process is Malthusian).

(C'.4) $\quad\quad \mu$ is supported by the lattice $\{0,1,2,...\}$ and by no sublattice.

Let us thus assume that (C.1)-(C'.4) are satisfied and let ξ_n, ξ_m, ν_n and μ_m be defined as in the continuous time case. We are now able to state

THEOREM (3.4)

Let N_k denote the total number of labels at time k except for the ancestral label. Then

(i) $\quad\quad$ For large values of k
$$E[N_k] \sim e^{\alpha k} \sum\limits_{j} e^{-\alpha j}\mu_m(\{j\})/\beta$$

(ii) $\quad\quad$ For large values of k
$$Var[N_k] \sim e^{2\alpha k} \frac{(\sum\limits_{j} e^{-\alpha j}\mu_m(\{j\}))^2 Var[\sum\limits_{j} e^{-\alpha j}\xi(\{j\})]}{\beta^2(1 - \sum\limits_{j} e^{-2\alpha j}\mu(\{j\}))}$$

(iii) $\quad\quad N_k/y_k \xrightarrow{\ p\ } \sum\limits_{j} e^{-\alpha j}\mu_m(\{j\})$ on $\{y_k \to \infty\}$, $k \to \infty$.

THEOREM (3.5)

If M_k denotes the number of present labels at k, except for the ancestral label, and $q_n(k) = P(Z_k^n = 0)$, where Z_k^n is the discrete time version of Z_t^n, then

(i) $\quad\quad E[M_k] \sim e^{\alpha k}(\sum\limits_{j} \mu_m(\{j\}))(\sum\limits_{h} e^{-\alpha h}(1 - q_n(h))/\beta$

$\quad\quad$ for large values of k.

(ii) $\quad\quad M_k/y_k \xrightarrow{\ p\ } (1 - e^{-\alpha})(\sum\limits_{j} e^{-\alpha j}\mu_m(\{j\})) \sum\limits_{h} e^{-\alpha h}(1 - q_n(h))$.

Let us now specialize further and turn our attention to the Galton-Watson case. (cf. Chapter One the notation of which will be used). In this case $\mu(j) = b$ if $j \geq 1$ and zero otherwise. If we let $\chi(j) = 1$ if $j = 0$, and zero otherwise, we get the process $Z_k :=$ Z_k^χ counting those born at k or equivalently the kth generation. Let the distribution of $B = \xi(\infty)$ be described by the p.g.f. $f(s) = \sum_j s^j p_j$ ($s \in [0, 1]$), where p_j is the probability of begetting j children and assume that each newly born individual is a mutant with the conditional probability θ, independently of everything else. We are then in the situation considered by [Pakes, 1984]. Let f_k denote the kth functional iterate of f. Then f_k is the p.g.f. of Z_k. Notice that conditionally on the event that an individual has begotten j offspring, the number of non-mutants among these will follow the binomial law, and thus the unconditional probability, P_k^θ, of giving birth to k non-mutant offspring is

$$P_k^\theta = \sum_j \binom{j}{k}(1 - \theta)^k \theta^{j-k} p_j.$$

This gives the following p.g.f. describing the reproduction of non-mutant offspring

$$g(s) = \sum_k P_k^\theta s^k = \sum_k \sum_j \binom{j}{k}(1 - \theta)^k \theta^{j-k} p_j s^k$$
$$= \sum_j p_j (\theta + (1 - \theta)s)^j$$
$$= f(\theta + (1 - \theta)s).$$

That is, $Z_k^{\mathbf{n}}$ will have $g_k(s)$ as p.g.f. Using well-known facts about Galton–Watson processes (cf. [Jagers, 1975]), we can write

$$E[Z_k] = b^k; \ Var[Z_k] = \sigma^2 b^{k-1}(b^k - 1)/(b - 1)$$

(recall that $b > 1$) where $\sigma^2 = Var[Z_1] = Var[B]$

$$E[Z_k^{\mathbf{n}}] = (1 - \theta)^k b^k$$

$$Var[Z_k^{\mathbf{n}}] = \frac{((1-\theta)^2\sigma^2 + b(1-\theta)\theta)b^{k-1}(1-\theta)^{k-1}(b^k(1-\theta)^k - 1)}{(b(1-\theta) - 1)}$$

if $b(1-\theta) \neq 1$, and

$$Var[Z_k^{\mathbf{n}}] = k((1-\theta)^2\sigma^2 + b(1-\theta)\theta) \ \text{if} \ b(1-\theta) = 1.$$

A matter that can be immediately settled in this light is the fate of the ancestral label. This will become extinct with the probability q_n such that

$$q_n = 1 \quad \text{if} \quad (1 - \theta)b \leq 1$$

$$q_n \text{ is the smallest root of } g(s) = s \quad \text{if} \quad (1 - \theta)b > 1.$$

Also in this case, some classical results can be used to say something about the time until eventual extinction of the ancestral label, and about its growth conditionally on not becoming extinct.

PROPOSITION (3.6)

Take $T = \inf\{j; Z_j^n = 0\}$, then

(i)   For $b(1 - \theta) = 1$ and $\sigma^2 < \infty$

$P(T > j) \sim 2/(j(\sigma^2(1 - \theta)^2 + b(1 - \theta)\theta))$, j large, and

$$P(Z_k^n/k \le u | Z_k^n > 0) \to 1 - \exp\left[\frac{-2u}{\sigma^2(1-\theta)^2 + b(1-\theta)\theta}\right], \ k \to \infty.$$

$$u \in R^+.$$

(ii)   For $b(1 - \theta) < 1$ and $\mathcal{O}(s) = \lim_{j \to \infty} (b(1-\theta))^j (1 - g_j(s))/(1-s)$

$$P(T > j) = 1 - g_j(0) \sim (b(1 - \theta))^j \mathcal{O}(0)$$

where $\mathcal{O}(0) > 0$ iff $\sum_j P_j^\theta j \log j < \infty$ and

$$\mathcal{O}(0) = 0 \text{ otherwise.}$$

It turns also out that some of the results in [Pakes, 1984] can be obtained in a rather confortable manner: Let M_k denote the number of present labels at k. In analogy with the continuous time case M_k can be written as the branching process counted by the random characteristic

$$\chi_x(j) = \begin{cases} 1 & \text{if } x \text{ is mutant and has at} \\ & \text{least one descendant with the same label} \\ & j \text{ generations later .} \\ 0 & \text{otherwise.} \end{cases}$$

Obviously $M_k = Z_k^\chi$. Before we formulate any results, notice that the equation defining the Malthusian parameter is

$$\sum_j e^{-\alpha j}\mu(\{j\}) = e^{-\alpha}\mu(\{1\}) = 1$$

and thus $e^{-\alpha j} = b^{-j}$ and that the mean age at childbearing, β, reduces to 1.

The following theorem illustrates the type of results that might be obtained:

THEOREM (3.7)

For large values of k,

(i) $E[M_k] \sim b^k \theta \sum_j b^{-j}(1 - g_j(0))$

(ii) $Var[M_k] \sim \dfrac{b^k \theta \sum_j b^{-j}(1 - g_j(0)))^2 \sigma^2}{b(b - 1)}$

(iii) $M_k/b^k \xrightarrow{\ p\ } \theta \sum_j b^{-j}(1 - g_j(0))W_\infty$, as $k \to \infty$

on the set of non-extinction.

(iv) $\qquad M_k/Z_k \xrightarrow{\ \ p\ \ } \theta \sum_j b^{-j}(1 - g_j(0)),$ as $k \to \infty.$

on the set of non-extinction.

PROOF

 (i), (iii) and (iv) follow from the results in APPENDIX B. For (ii) cf. the discussion at the end of §4.1. $\qquad\qquad\qquad\qquad\qquad\qquad\qquad\qquad$ □

Notice that $g_j(0)$ can be interpreted as the extinction probability within j generations of the ancestral process.

CHAPTER FOUR

THE LABEL PROCESS

An alternative approach to the study of some aspects of labelled branching processes is to consider the embedded unlabelled process formulated directly in terms of the labels themselves thought of as "individuals" in a population of labels. The resulting branching process, let us call it the label process, turns out to be completely characterized by the underlying one. This is not very surprising since nothing new is involved. In Chapter Ten we shall discuss the logical foundation of this and other similar embeddings. In §4.1 we present an attempt to actually construct the label process and to check its "good behaviour" in the sense of Chapter One. In the remaining sections we describe some aspects of the labelled population that are more conveniently formulated and studied in terms of the label process.

4.1. THE EMBEDDED LABEL PROCESS

In Chapter Ten we present a quite general embedding idea according to which individuals in a branching population can be grouped in such a way that the population of such groups behaves again as a branching population. The application of this idea to our labelled branching processes is obvious. The group of individuals carrying some particular label is seen as a generalized individual which can be identified with the label in question. Using the methods of Chapter Ten it can be proved that the population of such generalized individuals (the label process) is again a branching process. In the present chapter we concentrate our efforts on the investigation of the behaviour of this new process. But before doing so let us notice that if the underlying process is of the Bellman-Harris type, then the corresponding label process will be a general branching process in the meaning that the labels will beget (label-) children according to point processes specified by the life length distribution and the mutation mechanism. If the underlying process is a Galton-Watson process, then the corresponding label process will be be a general branching process but with lattice reproduction.

It is sraightforward to see that the reproduction of labels can be described by means of the following point process (in the sequel we write ξ', μ', etc for the reproduction point process, the intensity measure, etc of the label process):

$$\xi'(t) = \sum_{k=1}^{\infty} \sum_{x \in N^k} 1_{[\sigma_x, \infty)}(t) \prod_{j=1}^{k-1} \rho(x_j, \omega_{x_1, \ldots, x_{j-1}})$$
$$(1 - \rho(x_k, \omega_{x_{[1]}})),$$

and that the intensity measure of this point process is given by

$$\mu'(t) = E[\xi'(t)] = \sum_{k=0}^{\infty} \mu_n^{*k} * \mu_m(t).$$

In order to be able to deal with the label process, we have to check at least those among the assumption in Chapter One that ensure its "good behaviour" and the applicability of the basic convergence theorem. This is done in the following theorem.

THEOREM (4.1)

(i) If $\exists\ \alpha$ such that $\hat{\mu}(\alpha) = 1$, then $\hat{\mu}'(\alpha) = 1$.

(ii) Take $\beta' = \int_0^{\infty} u\, e^{-\alpha u} \mu'(du)$, then $\beta' = \beta/\hat{\mu}_m(\alpha)$.

(iii) $\mathrm{Var}[\hat{\xi}'(\alpha)] = \dfrac{\mathrm{Var}[\hat{\xi}(\alpha)]}{1 - \hat{\mu}_n(2\alpha)}.$

(iv) If $\exists\ \delta < \alpha$ such that $\hat{\mu}(\delta) < \infty$, then δ can always be taken so that $1 > \hat{\mu}_n(\delta) > \hat{\mu}_n(\alpha)$ in which case $\hat{\mu}'(\delta) < \infty$.

PROOF

(i) Let γ be such that $\hat{\mu}_n(\gamma) < 1$. Then $\hat{\mu}'(\gamma) = \int_0^{\infty} e^{-\gamma u} \mu_m * \nu_n(du) =$

$\hat{\mu}_m(\gamma)/(1 - \hat{\mu}_n(\gamma))$, where $\nu_n(u) = \sum_{k=0}^{\infty} \mu_n^{*k}(u)$. Since $\hat{\mu}(\alpha) = \hat{\mu}_n(\alpha) + \hat{\mu}_m(\alpha)$, it follows that $\hat{\mu}'(\alpha) = 1$, provided $\hat{\mu}_m(\alpha) > 0$. But otherwise $\mu' = \mu$.

(ii) By (i) it follows that $H(t) = \int_0^{t} e^{-\alpha u} \mu_m * \nu_n(du)$ is a probability distribution function with mean β'. But since $H(dt) = e^{-\alpha t} \mu_m * \nu_n(dt)$, the Laplace transform of H is

$$\hat{H}(\theta) = \int_0^{\infty} e^{-(\alpha+\theta)u} \mu_m * \nu_n(du)$$

$$= \hat{\mu}_m(\alpha+\theta)/(1 - \hat{\mu}_n(\alpha+\theta)).$$

Differentiation yields

$$\hat{H}'(0) = \frac{\hat{\mu}_m(\alpha) \int_0^{\infty} e^{-\alpha u} u\ \mu_m(du) + \hat{\mu}_m(\alpha) \int_0^{\infty} e^{-\alpha u} u\ \mu_n(du)}{\hat{\mu}_m(\alpha)^2}$$

$$= \beta/\hat{\mu}_m(\alpha).$$

(iii) It is not hard to realize that ξ' can be written in terms of the ancestral process as the general branching process which has ξ_n as reproduction point process and which is counted by a random characteristic $\chi(u) = \xi_m(u)$. Each individual with the same label as the ancestor contributes with the number of mutants among her offspring born up to age u (that is $\xi'(t) = Z^\chi_{n,t}$). This gives $\text{Var}[\hat{\xi}'(\alpha)] = \text{Var}[\int_0^\infty \alpha \, e^{-\alpha t} Z^\chi_{n,t} dt]$. Now the variance decomposition

$$\text{Var}[\int_0^\infty \alpha e^{-\alpha t} Z^\chi_{n,t} dt] = E[\text{Var}[\int_0^\infty \alpha e^{-\alpha t} Z^\chi_{n,t} dt \,|\, \mathscr{A}_0]]$$

$$+ \text{Var}[E[\int_0^\infty \alpha e^{-\alpha t} Z^\chi_{n,t} dt \,|\, \mathscr{A}_0]]$$

and the basic decomposition yield

$$\text{Var}[\hat{\xi}'(\alpha)] = E[\text{Var}[\int_0^\infty \alpha e^{-\alpha t} \xi_{m,0}(t) dt$$

$$+ \int_0^\infty \alpha \sum_{j \in N} Z^\chi_{n,t-\sigma}(j) dt \,|\, \mathscr{A}_0]]$$

$$+ \text{Var}[E[\int_0^\infty \alpha e^{-\alpha t} \xi_{m,0}(t) dt + \int_0^\infty \alpha e^{-\alpha t} \sum_{j \in N} Z^\chi_{n,t-\sigma_j}(j) dt \,|\, \mathscr{A}_0]].$$

But the first of the two terms at the right hand side of the above equality equals

$$E[\text{Var}[\int_0^\infty \alpha e^{-\alpha t} \sum_{j \in N} Z^\chi_{n,t-\sigma_j}(j) dt \,|\, \mathscr{A}_0]].$$

Since the daughter processes of the individuals in the first generation are conditionally independent, this also equals

$$E[\sum_{j \in N} \text{Var}[\int_0^\infty \alpha e^{-\alpha t} Z^\chi_{n,t-\sigma_j}(j) dt \,|\, \mathscr{A}_0]]$$

$$= E[\int_0^\infty \text{Var}[\int_0^t \alpha e^{-\alpha t} Z^\chi_{n,t-u} dt] \xi_n(du)].$$

Putting $s = t - u$ gives that this will again equal

$$E[\int_0^\infty e^{-2\alpha u} \text{Var}[\int_0^\infty \alpha e^{-\alpha s} Z^\chi_{n,s} ds] \xi_n(du)]$$

$$= \text{Var}[\hat{\xi}'(\alpha)] \hat{\mu}_n(2\alpha).$$

The second term in the variance decomposition of $\text{Var}[\hat{\xi}'(\alpha)]$ can be rewritten according to

$$\text{Var}[\hat{\xi}_m(\alpha) + \int_0^\infty \alpha e^{-\alpha t} \sum_{j \in N} E[Z^\chi_{n,t-\sigma_j}(j) \,|\, \mathscr{A}_0] dt]$$

$$= \text{Var}[\hat{\xi}_m(\alpha) + \int_0^\infty \alpha e^{-\alpha t} \int_0^t \nu_n * \mu_m(t-u) \xi_n(du) dt]$$

$$= \text{Var}[\hat{\xi}_m(\alpha) + \hat{\xi}_n(\alpha)] = \text{Var}[\hat{\xi}(\alpha)].$$

Summing the two terms yields

$$\mathrm{Var}[\hat\xi{}'(\alpha)] = \mathrm{Var}[\hat\xi{}'(\alpha)]\hat\mu_n(2\alpha) + \mathrm{Var}[\hat\xi(\alpha)].$$

iv)   First notice that

$$\hat\mu{}'(\theta) = \int_0^\infty e^{-\theta t}\mu_m * \nu_n(dt)$$

$$= \hat\mu_m(\theta)/(1 - \hat\mu_n(\theta)) = (\hat\mu(\theta) - \hat\mu_n(\theta))/(1 - \hat\mu_n(\theta)).$$

Obviously there is a $\delta < \alpha$ such that $\hat\mu(\delta) < \infty$ and $\hat\mu_n(\delta) < 1$, and thus

$$\hat\mu{}'(\delta) = (\hat\mu(\delta) - \hat\mu_n(\delta))/(1 - \hat\mu_n(\delta)) < \infty. \qquad \square$$

Observe also that μ' is non-lattice exactly when μ is non-lattice. Having specified the basic tools of the label process, we can now study the behaviour of subpopulations of labels counted by random characteristics. The asymptotics of such subpopulations are immediate consequences of the convergence of empirical distributions as in the basic convergence theorems in APPENDIX A. As an illustration of this we notice that the total population process

$$y'(t) = \#\{x \in I;\ x \text{ mutant and } \sigma_x \le t\}$$

satisfies

$$e^{-\alpha t}E[y'(t)] \to 1/\alpha\beta' = \hat\mu_m(\alpha)/\alpha\beta$$

which confirms what we have already shown in Chapter Three. But this is merely one simple application of the embedding idea. More interesting examples will be discussed in §4.2–4.3 (cf. also the proofs in Chapter Eight).

Before we close the present section, let us turn back to the number M_t of present labels and notice that the characteristic counting it was obviously non-individual, hence the non-applicability of Th. 2 in APPENDIX A. In terms of the label process, however, we get a branching process counted by an individual random characteristic and therefore

$$\mathrm{Var}[M_t] \simeq \frac{e^{2\alpha t}\hat\mu_m(\alpha)^2(1-\dot q(\alpha))^2\mathrm{Var}[\hat\xi{}'(\alpha)]}{\alpha^2\beta^2(1-\hat\mu(2\alpha))/(1-\hat\mu_n(2\alpha))}$$

$$= \frac{e^{2\alpha t}\hat\mu_m(\alpha)^2(1-\dot q(\alpha))\mathrm{Var}[\hat\xi(\alpha)]}{\alpha^2\beta^2(1-\hat\mu(2\alpha))}$$

where we use THEOREM (4.1) to identify $\mathrm{Var}[\hat\xi{}'(\alpha)]$ with $\mathrm{Var}[\hat\xi(\alpha)](1 - \hat\mu_n(2\alpha))$.

This means that provided the usual conditions on the underlying process are satisfied $e^{-2\alpha t}\mathrm{Var}[M_t]$ will converge.

4.2. THE OLDEST LABELS

Under this heading we consider the (asymptotic) behaviour of the process of the ages of the oldest labels as time passes. By the age of a label we simply mean the time elapsed since it arose by mutation. More precisely, we will show that this process becomes mixed Poisson in character, in the sense of weak convergence of point processes, as time increases. The technique used is the one described in [Jagers & Nerman, 1984b]. In that reference a distributional limit theorem is given concerning supercritical branching processes counted by (individual) random characteristics whose very definition may depend on absolute time. For ease of reference, this limit theorem is reproduced in APPENDIX A (TH. 7). The same paper contains a number of applications, one of which (APPLICATION C) is directly relevant for our purposes.

The reason why we must use such time dependent random characteristics and not characteristics of the type used so far is the following: We cannot give an a priori fixed age such that the ages of the oldest labels are scattered around this age at every moment of time. On the other hand it is possible to define a time dependent characteristic counting those oldest labels.

Recall from Chapter Two that T stands for the time until extinction of the ancestral process (i.e. $T = \inf\{u; Z_u^n = 0\}$) and that the tail probabilities are given by THEOREMS (2.1) and (2.2). Obviously T can be thought of as a natural life span of a label and $L(u) = P(T \leq u)$ can be identified as the life span distribution function. In what follows we discuss how the time dependent characteristic counting the oldest labels should be defined, i.e. we give the word old a precise meaning. The first step consists in finding an approximation of the backward time, c_t, starting at t which is such that the mean number of present labels older than c_t asymptotically equals one. Intuitively, c_t are non-negative numbers such that the expected number of new labels generated between time $t - c_t$ and time t with at least one living representative at time t is asymptotically equal to one. c_t has thus to satisfy

$$(4.3.1) \qquad \hat{\mu}_m(\alpha)e^{\alpha t} \int_{c_t}^{\infty} \alpha\, e^{-\alpha t}(1 - L(u))du/\alpha\beta \rightarrow 1,$$

as $t \rightarrow \infty$. When the ancestral process is subcritical, we have the approximation

$$1 - L(u) \sim c\, e^{\alpha_n u}$$

and (4.3.1) is satisfied for

$$(4.3.2) \qquad c_t = \frac{\alpha}{\alpha - \alpha_n} t + \frac{1}{\alpha - \alpha_n} \log \frac{c\, \hat{\mu}_m(\alpha)}{(\alpha - \alpha_n)\beta} \,, \quad \text{as } t \rightarrow \infty,$$

taken to define c_t.

Assume now that the ancestral process is critical. The corresponding approximation of $1 - L(u)$ is then given by

$$1 - L(u) \sim 2 \beta_n / \sigma_n^2 u$$

and the equation defining c_t becomes

$$\hat{\mu}_m(\alpha) e^{\alpha t} \int_{c_t}^{\infty} \frac{\alpha e^{-\alpha t} 2 \beta_n}{\sigma_n^2 u} \, du \to 1, \quad \text{as } t \to \infty.$$

We write this equation in the form

(4.3.3) $$k e^{\alpha t} \int_{\alpha c_t}^{\infty} \frac{e^{-\alpha u}}{u} \, du \to 1, \quad \text{as } t \to \infty$$

where $k = 2\hat{\mu}_m(\alpha)\beta_n / \beta \, \sigma_n^2$. The integral in this latter equation can be approximated by $e^{-\alpha c_t} / \alpha c_t$ (cf. [Abramovitz & Stegun, 1965] where such integrals are called the exponential integrals). We have now the simpler equation

(4.3.4) $$\frac{k \, e^{\alpha t} e^{-\alpha c_t}}{\alpha \, c_t} \to 1, \quad \text{as } t \to \infty$$

which is (asymptotically) satisfied when

(4.3.5) $$c_t = t - \frac{1}{\alpha} \log (\alpha t / k).$$

We take (4.3.5) as our definition of c_t in the case where the ancestral process is critical.

We can now formulate

PROPOSITION (4.2)

Let $\eta_t([a,b])$ be the process giving the number of present labels at t with ages in the age interval $c_t + [a,b]$ where $-\infty < a < b \leq \infty$. Then $\eta_t([a,b])$ converges as $t \to \infty$ in distribution towards a mixed Poisson r.v. with parameter

$$W_\infty(e^{-(\alpha-\alpha_n)a} - e^{-(\alpha-\alpha_n)b})$$

where the mixture coefficient W_∞ has the limiting distribution of $\alpha\beta' e^{-\alpha t} y'(t)$, and $\alpha_n = 0$ when the ancestral process is critical.

PROOF

Here we can rely on (TH.7) in APPENDIX A, the conditions of which are discussed in [Jagers & Nerman, 1984b] (p.64). The idea here is that the process $\eta_t([a,b])$ can be written as the branching process counted by the time dependent characteristic χ_t evaluated at $t' = t - c_t$ by

$$\chi_{t',x}(u) = \begin{cases} 1 \text{ if } x \text{ is alive at } t = t'+c_t \text{ and } u \in [a,b] \\ 0 \text{ otherwise} \end{cases}$$

Using (TH. 7) in APPENDIX A, it is seen that the process $\eta_t([a,b]) = Z_{t'}^{\chi_{t'}}$ converges to a r.v. whose characteristic function is

$$E[e^{W_\infty \hat{\mu}_m(\alpha)\hat{\psi}_\alpha(\theta)/\alpha\beta}]$$

where

$$\hat{\psi}_\alpha(\theta) = \lim_{t' \to \infty} e^{\alpha t'} \int_{-\infty}^{\infty} e^{-\alpha u}\{E[e^{i\theta\chi_{t'}(u)}] - 1\}du.$$

But

$$E[e^{i\theta\chi_{t'}(u)}] = e^{i\theta}P(T>u+c_t; u \in [a,b]) +$$

$$1 - P(T > u+c_t; u \in [a,b])$$

$$= (e^{i\theta} - 1)P(T > u+c_t; u \in [a,b]) + 1$$

which gives

$$\hat{\psi}_\alpha(\theta) = (e^{i\theta} - 1) \lim_{t' \to \infty} e^{\alpha t'} \int_a^b e^{-\alpha u}P(T > u+c_t)du$$

$$= (e^{i\theta} - 1) \lim_{t \to \infty} e^{\alpha t} \int_a^b e^{-\alpha(u+c_t)}P(T > u + c_t)du$$

$$= (e^{i\theta} - 1) \int_a^b \lim_{t \to \infty} e^{\alpha t}e^{-\alpha u}(1 - L(u))du$$

$$= \frac{(e^{i\theta} - 1)\alpha\beta(e^{-(\alpha-\alpha_n)a} - e^{-(\alpha-\alpha_n)b})}{\hat{\mu}_m(\alpha)},$$

where we use the fact that c_t solves equation (4.3.1) asymptotically. Substituting this in the above formula for the characteristic function of the limit variable yields the expression

$$E[(e^{e^{i\theta}} - 1)(e^{-(\alpha-\alpha_n)a} - e^{-(\alpha-\alpha_n)b})W_\infty]$$

(with W_∞ as above and $\alpha_n = 0$ in the critical case) which is recognized as the characteristic function of a mixed Poisson r.v. with the parameter

$$(e^{-(\alpha-\alpha_n)a} - e^{-(\alpha-\alpha_n)b})W_\infty. \qquad \square$$

The arguments of the above proof can be used to cover the following situation: Let S stand for the interval $(-\infty, +\infty)$. For every $t \in R^+$ and $I_1,..., I_n$ disjoint intervals of the form $(c_j, d_j] \subset S$, $j = 1,..., n$ take $t' = t - c_t$ and

$$\chi_{t',x}(u) = \begin{cases} 1 & \text{if } x \text{ is alive at } t = t'+c_t \text{ and } u \in \bigcup_{j=1}^{n} I_j. \\ \\ 0 & \text{otherwise} \end{cases}$$

Then $\eta_t(\bigcup_{j=1}^{n} I_j) = Z_t^{\chi_{t'}}$, giving the number of present labels at t with ages then in the

set $c_t + \bigcup_{j=1}^{n} I_j$, converges in distribution to a mixed Poisson r.v. with the parameter

$$W_\infty e(\bigcup_{j=1}^{n} I_j) = W_\infty \sum_{j=1}^{n} (e^{-(\alpha-\alpha_n)c_j} - e^{-(\alpha-\alpha_n)d_j})$$

when $I_j = (c_j, d_j] \cdot$ (where

$$e(A) = (\alpha - \alpha_n) \int_A e^{-(\alpha-\alpha_n)t} dt$$

is the exponential distribution measure with parameter $(\alpha-\alpha_n)$).

Now the above, together with the fact that Poisson processes are simple (admit no multiple points), imply that the conditions of (TH. A.1) in [Leadbetter et al, 1983] are satisfied and we can conclude.

THEOREM (4.3)

The point process η_t converges in distribution on the set of non-extinction towards a mixed Poisson point process with the directing measure e as described above. The convergence in the theorem takes place in the semi-weak topology, i.e. $\int_{-\infty}^{\infty} f(u)\eta_t(du)$ converges for bounded continuous functions f with a support bounded to the left towards $\int_{-\infty}^{\infty} f(u)\eta(du)$ in distribution.

As an immediate corollary to the above theorem we can formulate

COROLLARY (4.4)

With U_t = the age of the oldest label at t we have

$$P(U_t \leq c_t + s) = P(\eta_t([s,\infty] = 0) \to E[\exp(-W_\infty e^{-(\alpha-\alpha_n)s})],$$

$$s \in (-\infty, \infty].$$

Similary, the whole sequence of translated extreme ages can be shown to converge weakly towards the distribution of the extreme points in η.

Observe also that the theorem as formulated is distributional. Through a sequence of arguments, it is possible to "identify" the random component W_∞ in the limit with the actual normalized size of the underlying process, i.e. to show that $(\alpha\beta\, e^{-\alpha t}y_t,\, \eta_t)$ converges weakly to a two-dimensional object $(W_\infty,\, \eta)$ where the distribution of η given W_∞ is Poissonian as above. In a second step one can use continuous mapping to show that one can replace c_t by a y_t - determined age-normalizing translation to get a homogeneous Poisson limit of the randomly translated ages. The interested reader is referred to [Nerman, 1987] for a brief discussion.

4.3.   THE MOST FREQUENT LABELS

When the ancestral process is supercritical, we can use the same technique as in the preceding section to describe the asymptotics of the most frequent labels. As a biproduct, we are also able to describe the tail of the asymptotic distribution, P_j, of the number of labels with exactly j representatives. This distribution is intimately related to the famous Ewens sampling formula. Some of the arguments used in this section are similar to those in [Pakes, 1987].

Arguing as in the preceding section, we let n_t be such that the mean number of labels with more than n_t representatives at time t equals one, as $t \to \infty$. This leads to the following (asymptotic) equation:

$$\lim_{t\to\infty} \frac{\hat{\mu}_m(\alpha)}{\beta} e^{\alpha t} \int_0^\infty e^{-\alpha u}P(Z_u^n > n_t)du = 1.$$

The exact definition of n_t is given in the following proposition where we use the following notation: $L(u) = P(\lambda \le u)$ is the individual life span disribution function, $r = \alpha/\alpha_n$ and W_n is the limit random variable in the limit (in different senses)

$$\lim_{t\to\infty} e^{-\alpha_n t} Z_t^n = \frac{W_n}{\alpha_n \beta_n} (1 - \hat{L}(\alpha_n))$$

PROPOSITION (4.5)

Assume that $E[\hat{\xi}_n(\alpha_n)^{(r + \epsilon)}] < \infty$ for some $\epsilon > 0$ and that there exists a γ such that $\gamma < \alpha/(r + \epsilon)$ and $\sup_u e^{-\gamma(r + \epsilon)u} E[(Z_u^n)^{(r + \epsilon)}] < \infty$. Then the proper choice of n_t is $ce^{\alpha_n t}$ where

$$c = \left[\frac{r^2 \hat{\mu}_m(\alpha) E[W_n^r]}{\alpha \beta} \right]^{1/r} (1 - \hat{L}(\alpha_n)) / \alpha_n \beta_n .$$

PROOF

Take $K_t = \dfrac{\hat{\mu}_m(\alpha)}{\beta} e^{\alpha t} \int_0^\infty e^{-\alpha u} P(Z_u^n > ce^{\alpha_n t}) du$. Then K_t can also be written in

the form

$$K_t = \frac{\hat{\mu}_m(\alpha)}{\beta} \{ e^{\alpha t} \int_0^{t-t'} e^{-\alpha u} P(Z_u^n > ce^{\alpha_n t}) du + e^{\alpha t} \int_{t-t'}^\infty e^{-\alpha u} P(Z_u^n > ce^{\alpha_n t}) du \}.$$

We now omit the first term and substitute u by $y + t$ to get

$$K_t \geq \frac{\hat{\mu}_m(\alpha)}{\beta} e^{\alpha t} \int_{-t'}^\infty e^{-\alpha y} e^{-\alpha t} P(Z_{y+t}^n > ce^{\alpha_n t}) dy$$

which implies that

$$\liminf_t K_t \geq \frac{\hat{\mu}_m(\alpha)}{\beta} \int_{-t'}^\infty e^{-\alpha y} \lim_t P(e^{-\alpha_n(t+y)} Z_{y+t}^n > ce^{-\alpha_n y}) dy.$$

But $e^{-\alpha_n y} Z_y^n$ is known to converge in distribution towards $\dfrac{W_n}{\beta_n} \int_0^\infty e^{-\alpha_n u} (1 - L(u)) du$

and

$$\liminf_t K_t \geq \frac{\hat{\mu}_m(\alpha)}{\beta} \int_{-t'}^\infty e^{-\alpha y} P(W_n > \frac{c \alpha_n \beta_n e^{\alpha_n y}}{(1 - \hat{L}(\alpha_n))}) dy.$$

For ease of notation, we let $I_{t'}$ denote the expression on the right hand side of the
above ineqality. Turning back to K_t and using Markov's inequality we see that

$$P(Z_u^n > ce^{\alpha_n t}) \leq (\frac{1}{ce^{\alpha_n t}})^x E[(Z_u^n)^x],$$

$x \in R^+$, and

$$\limsup_t K_t \leq I_{t'} + \lim_t \frac{\hat{\mu}_m(\alpha)}{\beta c^x} e^{\alpha t} e^{-\alpha_n tx} \int_0^{t-t'} e^{-\alpha u} E[(Z_u^n)^x] du.$$

Taking $x = r + \epsilon$, $\epsilon > 0$ yields

$$\limsup_t K_t \leq I_{t'} + \lim_t \frac{\hat{\mu}_m(\alpha)(e^{\alpha_n \epsilon})^{-t}}{\beta\, c^{(r+\epsilon)}}\, t\text{-}t' \int_0^{t-t'} e^{-\alpha u} E[(Z_u^n)^{(r+\epsilon)}]\, du.$$

If γ satisfies the condition in the statement of the proposition, we can write

$$\alpha = (\alpha - \gamma r - \gamma\epsilon) + (\gamma r + \gamma\epsilon)$$

where $\alpha - \gamma r - \gamma\epsilon$ is strictly positive. This gives

$$\limsup_t K_t \leq$$

$$I_{t'} + \lim_t \frac{\hat{\mu}_m(\alpha)(e^{\alpha_n \epsilon})^{-t}}{\beta}\, t\text{-}t' \int_0^{t-t'} e^{-(\alpha - \gamma r - \gamma\epsilon)u}\, e^{-\gamma(r+\epsilon)u}\, E[(Z_u^n)^{(r+\epsilon)}]\, du$$

Since $e^{-\gamma u(r+\epsilon)} E[(Z_u^n)^{(r+\epsilon)}] < \infty$, dominated convergence implies that the limit on the right hand side of the last inequality will converge to zero at the same rate as $(e^{\alpha_n \epsilon})^{-t}$. We have proved that $\limsup_t K_t \leq I_{t'}$. Combined with $\liminf_t K_t \geq I_{t'}$, this yields

$$\lim_t K_t = \lim_{t'} I_{t'} = \frac{\hat{\mu}_m(\alpha)}{\beta} \int_{-\infty}^{\infty} e^{-\alpha y}\, P\left(W_n > \frac{c\alpha_n \beta_n e^{-\alpha_n y}}{(1 - \hat{L}(\alpha_n))}\right) dy.$$

Taking $k = c\alpha_n \beta_n /(1 - \hat{L}(\alpha_n))$ and $x = ke^{-\alpha_n y}$ gives

$$\lim_t K_t = \frac{\hat{\mu}_m(\alpha)}{\beta} \int_0^{\infty} \frac{x^{r-1}}{\alpha_n k^r}\, P(W_n > x)\, dx$$

$$= \frac{\hat{\mu}_m(\alpha)\, r^2}{\alpha\, \beta\, k^r} \int_0^{\infty} \frac{x^{r-1}}{r}\, P(W_n > x)\, dx$$

$$= \frac{\hat{\mu}_m(\alpha) \; r^2}{\alpha \; \beta \; k^r} \; E[W_n^r] = 1,$$

where the last step can be verified by inserting the proposed value of c, and the finiteness of $E[W_n^r]$ follows from the finiteness of $E[\hat{\xi}_n(\alpha_n)^{(r + \epsilon)}]$ (cf. [Asmussen & Hering, 1983]).   □

We are now ready to state the principal result of this section.

THEOREM (4.6)

Let n_t be defined as in PROPOSITION (4.5) and assume that the assumptions of that proposition are satisfied. Let further W_∞ be the limit variable in PROPOSITION (4.2). Then the number of present labels with more than n_t representives at time t converges, as $t \to \infty$, in distribution towards a mixed Poisson random variable with parameter W_∞/α.

PROOF

As in the proof of PROPOSITION (4.2) we can rely on the main theorem in [Jagers & Nerman, 1984b]. All we have to do is to consider the branching process, $Z_t^{\chi_t}$, counted by the time dependent random characteristic

$$\chi_t(u) = \begin{cases} 1 & \text{if} \quad Z_u^n > c e^{\alpha_n t} \\ 0 & \text{otherwise.} \end{cases}$$

According to the above mentioned reference, Z_t^χ converges in distribution towards a random variable whose characteristic function, evaluated at θ, is of the form

$$E[e^{W_\infty \hat{\mu}_m(\alpha) \hat{\psi}_\alpha(\theta)/\alpha\beta}]$$

where

$$\hat{\psi}_\alpha(\theta) = \lim_{t\to\infty} \int_{-\infty}^{\infty} e^{\alpha t}(\varphi_{tu}(\theta) - 1)e^{-\alpha u} du$$

and

$$\varphi_{tu}(\theta) = E[e^{i\theta\chi_t(u)}].$$

But in our case

$$\varphi_{tu}(\theta) = e^{i\theta} P(Z_u^n > ce^{\alpha_n t}) + 1 - P(Z_u^n > ce^{\alpha_n t})$$

and

$$\hat{\psi}_\alpha(\theta) = \lim_{t \to \infty} e^{\alpha t} \int_0^\infty e^{-\alpha u}(e^{i\theta} - 1)\, P(Z_u^n > ce^{\alpha_n t})du$$

which by PROPOSITION (4.5) equals $\beta(e^{i\theta} - 1)/\hat{\mu}_m(\alpha)$.   □

Let us now turn our attention to the asymptotic proportions, P_j, of labels represented by exactly j living individuals, $j = 1,2,\dots$. With

M_t^j = the number of labels with exactly j living representives at time t

and

M_t = the number of present labels at time t

such proportions can be studied using the limits, as t tends to infinity, of the empirical ratios M_t^j/M_t. Since M_t^j and M_t can be written as branching processes counted by random characteristics we can use the same technique as in Chapter Three to get the following result

PROPOSITION (4.7)

The asymptotic proportion of labels with exactly j representives is

$$P_j = \frac{\alpha}{1 - \hat{q}_n(\alpha)} \int_0^\infty e^{-\alpha u}\, P(Z_u^n = j)du$$

where $q_n(t) = P(Z_t^n = 0)$.

Unfortunately it is not possible to be more specific about P_j quite generally. Notice, however, that the tail of this distribution is of the form

$$Q(k) = \sum_{j>k} P_j = \frac{\alpha}{1 - \hat{q}_n(\alpha)} \int_0^\infty e^{-\alpha u}\, P(Z_u^n > k)du.$$

PROPOSITION (4.5) has now the following immediate corollary which gives a qualitative description of the tail of the above distribution.

COROLLARY (4.8)

Under the conditions of PROPOSITION (4.5)

$$\frac{e^{\alpha t}(1 - \hat{q}_n(\alpha))\hat{\mu}_m(\alpha)}{\alpha \ \beta} Q(n_t) \to 1,$$

as $t \to \infty$.

CHAPTER FIVE

THE LIMITING STABLE CASE

A well established fact about supercritical branching population processes is that they either become extinct or explode at an exponential rate. Still it is possible to imagine some kind of asymptotic stability even in this case, namely that of the composition, in several senses, of a population according to some appropriate limit composition laws, even though the population size continues to grow. This idea has been exploited by many authors. A very useful way of thinking about such composition laws is to view them as related to some randomly sampled individual chosen in some convenient manner from a very old branching population (see below). Probabilistic statements concerning different aspects of the pedigree of this sampled individual can then be done using the probability measure $\tilde{P}$ the stable pedigree measure, which in a sense summarizes the above mentioned limit laws, and which is induced by the sampling mechanism and the life law. In the present chapter we make some definitions and introduce some notation to be used in the coming chapters as a straightforward tool in the study of retrospective aspects of labelled branching processes. The whole chapter is more or less based on [Jagers & Nerman 1984a] and [Nerman & Jagers, 1984].

5.1. THE STABLE PEDIGREE SPACE

Consider a well behaved supercritical branching process in the sense of Chapter One. Assume that a randomly sampled individual (RSI) is chosen among all individuals either dead or alive at time t (in the sequel RSI will always refer to sampling in this maximal category). In biological applications, however, it is more useful to have results concerning a randomly sampled individual chosen among those alive at the sampling time point. As will be described in §5.3 the results corresponding to sampling among those alive are obtained by a slight modification of those valid in the case of an RSI.

Imagine the pedigree of the RSI as the object describing the relations between the individuals in the process from her point of view. Such an object can also be thought of as containing the lives of the concerned individuals as well as some time coordinate. It turns out to be convenient to work with the enlarged space of individuals $J = (Z^-) \times I = \bigcup_{j=0}^{\infty} (\{-j\} \times I)$, where $-0 = 0$ stands for the RSI herself, $-j$ is her jth ancestor, and $\{-j\} \times I$ is the space of individuals stemming from her jth ancestor except for $-(j-1)$ and her progeny. The following notation will be used in the sequel

$$J_n = \bigcup_{j=0}^{n} \{-j\} \times I$$

i_k, $k \in N$ is $-(k-1)$'s birth order among the offspring of $-k$.

$(i_k)_{k \in N}$ describes the ancestry of the RSI.

$$\tilde{\Omega} = \Omega^J \times N^\infty$$
$$\tilde{\Omega}_n = \Omega^{J_n} \times N^n.$$

The pedigree space can be formally defined as $\tilde{\Omega} \times R^+$ and a pedigree as an element of the pedigree space and thus of the form $(\tilde{\omega}, a)$ where $\tilde{\omega}$ denotes a typical element in $\tilde{\Omega}$ and a is the age of the RSI at the sampling time point. Clearly $(\tilde{\omega}, a) = (\{\omega_x\}_{x \in J}, (i_k)_{k \in N}, a)$.

Let us now redefine the sampling time point as being the new time origin. The birth moment of the RSI relative to this new time origin is $\tilde{\sigma}_0(\tilde{\omega}, a) = -a$, and those of her direct ancestors can be defined according to $\tilde{\sigma}_{-k} = \tilde{\sigma}_{-(k-1)} - \tau(i_k, \omega_{-k})$. The birth moments of the other individuals are described by:

$$\tilde{\sigma}_{-kj} = \begin{cases} \tilde{\sigma}_{-k} + \tau(j, \omega_{-k}) & \text{if } j < i_k \\ \tilde{\sigma}_{-k} + \tau(j+1, \omega_{-k}) & \text{if } j \geq i_k \end{cases}$$

and

$$\tilde{\sigma}_{xj} = \tilde{\sigma}_x + \tau(j, \omega_k), \quad x \in J \setminus \{-1, -2, \ldots\}, \; j \in N.$$

To summarize, the sampling time point is now taken to be the new time origin and the RSI (-0 or simply 0 in the new notation) is born at $-a$ and is the i_1th daughter of her mother, -1. At her turn -1 is the i_2th daughter of -2 and was born at $-(a + \tau(i_1, \omega_{-1}))$ etc.

The pedigree space is equipped with the product σ-algebra $\mathscr{A}^J \times \mathscr{N} \times \mathscr{B}$ where $\mathscr{N}$ stands for the product σ-algebra $2^N \times 2^N \times \ldots$ and $\mathscr{B}$ for the usual family of Borel sets in R^+. The relation between the old measurable space $(\Omega^I, \mathscr{A}^I)$ and the new one $(\tilde{\Omega} \times R^+, \mathscr{A}^J \times \mathscr{N} \times \mathscr{B})$ is best understood in terms of the mappings $\pi_x : I \longmapsto J_n$, $x \in N^n$, $n \in N$ centering the population around x (i.e. rendering her the RSI while keeping all family relations untouched). This means that $\pi_x(x) = 0$ and for all y and z in I, there are unique $\pi_x(y)$ and $\pi_x(z)$ such that $\pi_x(y)$ is related to $\pi_x(z)$ in the same way as y is related to z. To this mapping there corresponds an operator $\Pi_x : \Omega^I \to \tilde{\Omega}$ transforming the lives in the appropriate way: $\Pi_x((\omega_y)_{y \in I}) = ((\omega_{\pi_x^{-1}(y)})_{y \in J}, x_\infty)$, (for x_∞ see below). Since the pedigree space is clearly too large (containing pedigrees extending infinitely many steps in the past of the process), the question arises of how to define the coordinates lying beyond the generation numbers of the actual randomly sampled individuals (i.e. $\omega_{\pi_x^{-1}(y)}$, y

$\in \{-(n(x) + k), k = 1,2,..., x \in I\}$). It turns out (cf. [Nerman & Jagers, 1984]) that those can be chosen by some arbitrary convention. This means that the expression $((\omega_{\pi^{-1}(y)})_{y \in J}, x_\infty)$ is to be interpreted according to: $\omega_{\pi_x^{-1}(y)}$ are arbitrary elements whenever $y \in J \backslash J_n$ and $x_\infty = (x_n, x_{n-1},...,x_1,i_{n+1},...)$ where $x_j = i_{n-j+1}$ and $(i_{n+k})_{k=1}^\infty$ are arbitrary fixed elements of N^∞. In these terms the pedigree of an individual $x \in I$ at some point, t, of time can be written as $(\Pi_x((\omega_y)_{y \in I}), t - \sigma_x)$. This provides us with a useful tool for describing functionals of the pedigree of the RSI as mappings $\varphi: \tilde{\Omega} \times R^+ \longrightarrow R^+$. Examples of such functionals will be given in Chapter Six as well as in later chapters.

5.2.   THE STABLE PEDIGREE MEASURE

When dealing with functionals of the pedigree of the RSI, at some late time point t, it would be nice (and time saving) to make probabilistic statements directly without having to define everything in terms of averages over the population. In other words it would be useful to be able to avoid going the long way using random characteristics and the basic convergence theorems and to work directly on the probability space

$$(\tilde{\Omega} \times R^+, \mathscr{A}^J \times \mathscr{N} \times \mathscr{B}, \tilde{P})$$

for some probability measure $\tilde{P}$ induced on the pedigree space by the population law and the sampling mechanism. In the following, we will introduce the appropriate probability measure $\tilde{P}$ on the space of pedigrees and mention some facts about its nature for ease of reference.

$\tilde{P}$ can be said to arise as the limit of empirical pedigree measures induced by sampling of an individual among those born during the pre–t history of the original branching process (i.e. a measure giving mass $1/y_t$ to each pedigree pertaining to some individual born before t, dead or alive, y_t being the total number of such individuals). Sampling among those alive is discussed in §5.3.

According to $\tilde{P}$ the age, a, of the RSI is exponentially (α) distributed and the lives in the set $\{\omega_x; x \in J$ but $x \neq -1, -2,...\}$ are i.i.d. following the life law P and are also independent of a and of the lives of the ancestors. Those at their turn are i.i.d. on the naturally extended space $\Omega \times N$ giving also information about the next step in the lineage to the RSI (i.e. i_k, $k \in N$ which themselves form an i.i.d. sequence). $\tilde{P}$ can thus be characterized by these independencies and the margins

$$\tilde{P}(\omega_x \in A) = P(\omega_x \in A) \text{ if } x \neq -1, -2, \ldots$$

$$\tilde{P}(\omega_{-k} \in A, i_k = j) = E[e^{-\alpha\tau(j)}; A], k \in N.$$

An immediate consequence of this is

$$\tilde{P}(\omega_{-k} \in A) = \sum_{j=1}^{\infty} \tilde{P}(\omega_{-k} \in A; i_k = j)$$

$$= \sum_{j=1}^{\infty} E[e^{-\alpha\tau(j)}; A]$$

$$= E[\hat{\xi}(\alpha); A].$$

Accordingly, the expectation with respect to $\tilde{P}$, denoted by $\tilde{E}$, will be such that

$$\tilde{E}[\eta(\omega_x)] = E[\eta] \qquad \text{if } x \neq -1, -2, \ldots$$

$$\tilde{E}[\eta(\omega_{-k})] = E[\eta\hat{\xi}(\alpha)], k \in N.$$

In later sections we will be concerned with measurable functions φ defined on

$$\tilde{\Omega} \times R^+, \quad \mathscr{A}^J \times \mathscr{N} \times \mathscr{B}).$$

In order to be able to deal with these, we define $\{\mathscr{A}_n\}$ as the sequence of σ–algebras generated by the projections

$$p_n \quad \tilde{\Omega} \longrightarrow \tilde{\Omega}_n$$

and consider the events $\mathscr{A}_n$ given by

$$\mathscr{A}_n = \{(\tilde{\omega}, a) \in \tilde{\Omega} \times R^+; \varphi(\tilde{\omega}, a) \text{ is determined by } (p_n\tilde{\omega}, a)\}.$$

To such events corresponds stopping times

$$\nu_\varphi(\tilde{\omega}, a) = \inf\{n; (\tilde{\omega}, a) \in A_n\}$$

with respect to the sequence $\{\mathscr{A}_n\}$ generated by the projections onto $\tilde{\Omega}_n \times R^+$,

$n - 0, 1, 2, \ldots$. The existence of such "minimal" stopping times can be argumented for using Zorns lemma. Notice now that, with φ being a functional of the pedigree of the RSI, if

$$\varphi_k(\tilde{\omega}, a) = 1_{\{k\}}(\nu_\varphi(\tilde{\omega}, a))\varphi(\tilde{\omega}, a)$$

for $a \geq 0$ and $k \in \{0, 1, 2, \ldots\}$ and φ_k defined for an arbitrary individual x ($\in I$, $n(x) \geq k$) as a function of her pedigree, in which case it takes the form $\varphi_k(\Pi_x, t - \sigma_x)$, then summation over all individuals defines

$$Z_t^\varphi = \sum_{x \in I} \sum_{k=0}^{n(x)} \varphi_k(\Pi_x, t - \sigma_x)$$

$$= \sum_{n=0}^{\infty} \sum_{x \in N^n} \sum_{k=0}^{n} \varphi_k(\Pi_x, t - \sigma_x).$$

Now using the usual resummation trick gives

$$Z_t^\varphi = \sum_{y \in I} \sum_{n=0}^{\infty} \sum_{x \in N^n} \varphi_n(\Pi_x \circ S_y, t - \sigma_y - \sigma_x \circ S_y).$$

Taking

$$\chi_x(u) = \sum_{x \in I} \varphi_{n(x)}(\Pi_x, u)$$

we can write

$$Z_t^\varphi = \sum_{y \in I} \chi_y(t - \sigma_y) = Z_t^\chi$$

with χ depending solely on the daughter processes of the counted individuals. Notice that we have avoided the arbitrary coordinates in Π_x by summing the φ_k values only up to x's generation number $n(x)$. In this case the usual convergence theorems apply (under suitable conditions on χ) on Z_t^χ, and we get

$$\frac{Z_t^\chi}{y_t} \longrightarrow E[\hat{\chi}(\alpha)]$$

in several senses on the set of non–extinction $\{y_t \longrightarrow \infty\}$. $E[\hat{\chi}(\alpha)]$ can now be interpreted as the mean φ–value of an RSI from an old (stable?) population, if we make the additional assumption that to the arbitrary coordinates in Π_x correspond zero φ–contributions. In [Nerman & Jagers, 1984] it is shown that the limit is exactly equivalent to the expectation of φ with respect to $\tilde{P}$ for the more general setting where

$\varphi : \tilde{\Omega} \times R^+ \longrightarrow R^+$ is any measurable function of the form $\varphi = \sum_{n=0}^{\infty} \varphi_n$ $\tilde{P}$–a.s. where

$\varphi_n \geq 0$ is $\mathscr{A}_n$–measurable, not necessarily defined as above.

This seems to suggest two alternative ways of dealing with the functionals, φ, of the pedigree of an RSI: The first is to use the stable pedigree measure directly and calculate $\tilde{E}[\varphi]$. The second is to define the appropriate random characteristic, χ, and consider the convergence of the empirical ratio $Z_t^\chi / y_t \longrightarrow E[\hat{\chi}(\alpha)]$ in different meanings. Applying the latter method means that a number of (sometimes non–trivial) resummations have to be done and that the conditions on χ have to be checked. We refer the reader to TH.6 in APPENDIX A for the exact conditions required in order for the first of the above methods to apply. The equivalence of the limits derived by the two methods is given in the following theorem which we borrow from the above reference:

THEOREM (5.1)

If $\varphi : \tilde{\Omega} \times R^+ \longrightarrow R^+$ can be written in the form $\varphi = \sum_{n=0}^{\infty} \varphi_n$ $\tilde{P}$–a.s. where

$\varphi_n \geq 0$ is $\mathscr{A}_n$–measurable, then taking

$$\chi(a) = \sum_{n=0}^{\infty} \sum_{x \in N^n} \varphi_n(\Pi_x, a - \sigma_x)$$

we have

$$E[\hat{\chi}(\alpha)] = \sum_{n=0}^{\infty} \sum_{x \in N^n} E[e^{-\alpha\sigma_x} \hat{\varphi}_n(\Pi_x, \alpha)] = \tilde{E}[\varphi].$$

We will also need the following lemma from [Jagers & Nerman, 1984a]

LEMMA (5.2)

Let $\varphi : \tilde{\Omega} \times R^+ \to R^+$ be measurable and determined by $\nu \leq$ some $n \in N$. Then

$$\tilde{E}[\varphi] = \sum_{x \in N^n} E[e^{-\alpha\sigma_x} \hat{\varphi}(\Pi_x, \alpha)].$$

5.3. SAMPLING AMONG LIVING INDIVIDUALS

Assume that everything is as above except that we choose our randomly sampled individual among all those alive at the sampling time point. The corresponding stable pedigree measure is then to be modified in the following way:

Expressions containing sums of φ-values have to be completed by adding indicator functions $1(\lambda_x > t - \sigma_x)$ where λ_x is x's life span. This gives

$$z_t^{\varphi} = \sum_{x \in I} \varphi(\pi_x, t - \sigma_x) 1(\lambda_x > t - \sigma_x).$$

The asymptotic limit in probability on $\{y_t \to \infty\}$ of the average φ-value becomes

$$\tilde{E}[\varphi | \lambda_0 > a] = \tilde{E}[\varphi; \lambda_0 > a] / \tilde{P}(\lambda_0 > a)$$

where a is the age of the randomly sampled individual at the sampling time point. Since in this case, the event of being alive at the sampling time point only depends on the life of the sampled individual, the stable pedigree measure is only modified in that the age and the life of the randomly sampled individual are no longer independent. As to the margins of other individuals, they are not affected by the conditioning. The same remains valid for sampling among individuals having some other "individual" property. In the sequel, the stable pedigree measure obtained when the sampling is restricted to those alive will be denoted by $\tilde{P}_\ell$ and $\tilde{E}_\ell$ will be the corresponding expectation.

CHAPTER SIX

THE PROCESS OF MUTANT ANCESTORS

Under this heading we will be concerned by some applications of what can be called the stable pedigree calculus. By this we mean that we consider some functionals of the pedigree of a randomly sampled individual (from a very old well behaved branching process) written in the terminology of the preceding chapter. Those functionals are such that the same results are valid under $\tilde{P}$ as under $\tilde{P}_\ell$ (see Chapter Five). For that reason it is motivated to call the randomly sampled individual RSI according to the convention of the preceding chapter. Occasionally, the precise limit theorems formulated in terms of the underlying process will be given.

More exactly our intention is to study the process describing the birthtimes of the mutant ancestors of the RSI.

6.1. THE AGE OF THE LABEL OF AN RSI

Let us, to begin with, focus our attention on the age of the label of an RSI. We will define this age as the length (in generations or in real time) of the time interval ending at the birth moment of the RSI and starting by the birth moment of her last mutant ancestor. Notice that the above definition of the age of a label relieves us from the trivial complication of adding the RSI's age everywhere in our formulas.

The study of various aspects of the age of a label has been, partially, the subject of many research papers. Examples are [Hoppe, 1984] and [Watterson, 1976].

Let $\nu = \nu(\tilde{\omega},a)$ stand for the generation age of the label of the RSI and take

$$A_n = \{(\tilde{\omega},a) \in \tilde{\Omega} \times R^+; \text{ at least one } \rho(i_k,\omega_{-k}) = 0,$$

$$k = 1,...,n\}.$$

Then clearly $\nu(\tilde{\omega},a) = \inf\{n;(\tilde{\omega},a) \in A_n\}$. In these terms the event $\{\nu = n\}$ will mean that the last mutant ancestor of the RSI is the i_nth daughter of $-n$ or equivalently $-(n-1)$.

PROPOSITION (6.1)

Under both $\tilde{P}$ and $\tilde{P}_\ell$, ν is geometrically distributed with parameter $\hat{\mu}_m(\alpha)$.

PROOF

Since $\hat{\mu}_m(\alpha) + \hat{\mu}_n(\alpha) = 1$, what we in fact need to prove is

$$\tilde{P}(\nu{=}k) = \hat{\mu}_m(\alpha)\hat{\mu}_n(\alpha)^{k-1}, \ k \in N.$$

But

$$\tilde{P}(\nu=k) = \tilde{P}(\rho(i_1,\omega_{-1}) = ... = \rho(i_{k-1},\omega_{-(k-1)})$$
$$= 1, \rho(i_k,\omega_{-k}) = 0)$$

and $\{\rho(i_k,\omega_{-k})\}_{k\in N}$ is obviously an i.i.d. sequence of $0-1$ random variables under $\tilde{P}$. It is thus enough to show that $\tilde{P}(\rho(i_k,\omega_{-k}) = 1) = \hat{\mu}_n(\alpha)$. But (cf. the previous chapter)

$$\tilde{P}(\rho(i_k,\omega_{-k}) = 1) = \sum_{j\in N} \tilde{P}(\rho(i_k,\omega_{-k}) = 1, i_k = j)$$

$$= \sum_{j\in N} E[e^{-\alpha\tau(j)};\rho(j) = 1] = E[\int_0^\infty e^{-\alpha u}\xi_n(du)] = \hat{\mu}_n(\alpha)$$

We now show that the statement of the proposition remains true under $\tilde{P}_\ell$. Let A stand for the event that our randomly sampled indivudual, 0, is alive at the sampling time point. Then the probability that 0 is a mutant given A is again

$$\tilde{P}_\ell(\rho(i_1, \omega_{-1}) = 0) = \tilde{P}(\rho(i_1, \omega_{-1}) = 0|A) = \hat{\mu}_m(\alpha)$$

due to the independence of a, ω_0 and (ω_{-1}, i_1) under $\tilde{P}$ which makes the events A and $\{\rho(i_1, \omega_{-1}) = 0\}$ independent. More generally, it is similarly seen that

$$\tilde{P}_\ell(\rho(i_k, \omega_{-k}) = 0) = \tilde{P}(\rho(i_k, \omega_{-k}) = 0|A) = \hat{\mu}_m(\alpha). \qquad \square$$

Observe that the mean and variance of ν are $\tilde{E}[\nu] = 1/\hat{\mu}_m(\alpha)$ and $\tilde{Var}[\nu] = \hat{\mu}_n(\alpha)/\hat{\mu}_m(\alpha)^2$.

NOTES

- As already mentioned, the precise limit theorems can be formulated in terms of suitable random characteristics, χ, such that Z_t^χ/y_t converges towards the probabilities $\tilde{P}(\nu=k)$, the mean $\tilde{E}[\nu]$ or the variance $\tilde{Var}[\nu]$. But it turns out that much more work is then needed in order to achieve the above simple result. In the case where we sample among living individuals, the same remains true if the total population count, y_t is replaced by Z_t, the process counting those still alive at t.

- In the case where the mutation mechanism is such that a new born child is a mutant with probability θ independently of everything else, then this applies to ancestors as well

$$\hat{\mu}_m(\alpha) = \int_0^\infty e^{-\alpha t}E[\xi_m(dt)] = \theta.$$

- It should be noticed that $\tilde{P}$(the RSI is mutant) $= \hat{\mu}_m(\alpha)$. This quantity is sometimes used to measure the evolution rate of the population.

— As was mentioned in the beginning of this chapter, the functionals we study are such that replacing $\tilde{P}$ by $\tilde{P}_\ell$ does not affect the results as can be seen by similar arguments as in the proof of PROPOSITION (5.1). For that reason only results based on $\tilde{P}$ will be formulated in what remains of this chapter.

Let $T \; (= T(\tilde{\omega},a))$ stand for the real time age of the label of the RSI as discussed above. Clearly

$$T = \tau_1 + ... + \tau_{\nu-1}$$

where

$$\tau_k = \tau(i_k,\omega_{-k}), \quad k \in N$$

and

$$T = 0 \text{ if } \nu = 1.$$

Typically we have a situation similar to that of figure 6.1

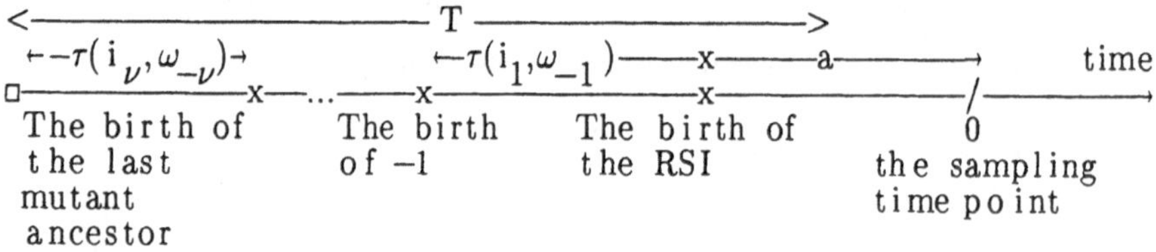

Fig. 6.1

PROPOSITION (6.2)

With $\mu_\alpha^n(t) = \int_0^t e^{-\alpha u} \mu_n(du)$ we have

(a) $\quad \tilde{P}(T \le t; \nu = k) = (\mu_\alpha^n)^{*k}(t) \, \hat{\mu}_m(\alpha)$

(b) $\quad \tilde{P}(T \le t) = \sum_{k=0}^{\infty} (\mu_\alpha^n)^{*k}(t) \, \hat{\mu}_m(\alpha)$

(c) $\quad \tilde{E}[T] = B_n/\hat{\mu}_m(\alpha)$

where $B_n = \int_0^\infty u e^{-\alpha u} \mu_n(du)$

PROOF

Assertion (b) follows directly from (a), for which PROPOSITION (6.1) yields

$$\tilde{P}(T \le t; \nu = k) = \tilde{P}(T \le t | \nu = k)\tilde{P}(\nu = k)$$
$$= \tilde{P}(T \le t \, | \, \rho(i,\omega_{-1}) = ... = \rho(i_{k-1},\omega_{-k+1}) = 1; \, \rho(i_k,\omega_{-k})$$
$$= 0) \, \hat{\mu}_m(\alpha)\hat{\mu}_n(\alpha)^{k-1}$$

$$= \tilde{P}(\tau_1 + \ldots + \tau_{k-1} \leq t \,|\, 0, -1, \ldots, -k+1 \text{ are non mutants while}$$

$$-k \text{ is mutant})\hat{\mu}_{\mathbf{m}}(\alpha)\hat{\mu}_{\mathbf{n}}(\alpha)^{k-1}.$$

But $\quad \tilde{P}(\tau(i_j, \omega_{-j}) \leq u \,|\, \rho(i_j, \omega_{-j}) = 1)$

$$= \tilde{P}(\tau(i_j, \omega_{-j}) \leq u; \rho(i_j, \omega_{-j}) = 1) / \tilde{P}(\rho(i_j, \omega_{-j}) = 1)$$

$$= \mu_{\alpha}^{\mathbf{n}}(u) / \hat{\mu}_{\mathbf{n}}(\alpha).$$

By independence

$$\tilde{P}(T \leq t; \nu = k) = (\mu_{\alpha}^{\mathbf{n}})^{*k-1}(t)\hat{\mu}_{\mathbf{m}}(\alpha)\hat{\mu}_{\mathbf{n}}(\alpha)^{k-1} / \hat{\mu}_{\mathbf{n}}(\alpha)^{k-1}$$

$$= (\mu_{\alpha}^{\mathbf{n}})^{*k-1}(t)\hat{\mu}_{\mathbf{m}}(\alpha).$$

(c)   Since T has the form

$$T = \tau_1 + \ldots + \tau_{\nu-1}$$

we have

$$\tilde{E}[T] = \tilde{E}\left[\sum_{i=1}^{\nu-1} \tau_i\right]$$

$$= \tilde{E}\left[\tilde{E}\left[\sum_{i=1}^{\nu-1} \tau_i \,|\, \nu\right]\right]$$

$$= \tilde{E}[(\nu-1)\tilde{E}[\tau_1]]$$

$$= \tilde{E}[\tau_1]\tilde{E}[\nu-1]$$

$$= \int_0^\infty u e^{-\alpha u} \mu_{\mathbf{n}}(du) \,/\, \hat{\mu}_{\mathbf{m}}(\alpha)$$

$$= B_{\mathbf{n}} \,/\, \hat{\mu}_{\mathbf{m}}(\alpha) \qquad\qquad\qquad \square$$

6.2.   A RENEWAL PROCESS

Since a large number of mutations is expected to have occured in the past family history of the RSI, the question: What is likely to have happened in her lineage during, say, the last t time units before her birth? almost poses itself.

To be able to make precise statements about the remote past of the process, we shall now consider the backward process of the birthtimes of the mutant ancestors of the RSI. Also in this case, the stable pedigree calculus turns out to be suitable. As pointed out in [Nerman & Jagers, 1984], the product form of the stable pedigree measure $\tilde{P}$ (and the same is valid for $\tilde{P}_\ell$) implies, in the usual "unlabelled" case, that the sequence $\{-\tilde{\sigma}_{-k}\}_{k=0}^\infty$ defines a (delayed) renewal process with interarrival time distribution function

$\mu_\alpha(t) = \int_0^t e^{-\dot{\alpha}u}\mu(du)$. In the present labelled case we can still obtain a similar renewal process. To see this define $X(t)$ as the number of (birth moments pertaining to) mutant ancestors of the RSI in the interval $[-a,-a-t]$. That $X(t)$ is a renewal process follows from the fact that the ancestors can be classified as mutants or non—mutants while the successive time spans between their birth times still define a renewal process. By independence the sequence $\{(\rho_{-k},\tilde{\sigma}_{-k}) , k = 1, 2 ,...\}$ satisfies

$$\tilde{P}(\rho_{-(k+1)} = j, -\tilde{\sigma}_{-(k+1)} + \tilde{\sigma}_{-k} \leq u \,|\rho_0,...,\rho_{-k},-\tilde{\sigma}_0,...,\tilde{\sigma}_{-k})$$
$$= \tilde{P}(\rho_{-(k+1)} = j, \tau(i_k,\omega_{-k}) \leq u|\rho_{-k})$$

for all $j \in \{0,1\}$, $k \in N$ and $u \in R^+$, and is thus a very simple Markov renewal process. A known fact about Markov renewal processes is that if we fix the state j (in this case 0) and consider the successive $-\tilde{\sigma}_{-k}$ such that $\rho_{-k} = j$ (i.e. 0) then those successive $-\tilde{\sigma}_{-k}$ define a renewal process. This describes how $X(s)$ arises as the (delayed) renewal process of (the birth times of) the mutant ancestors of the RSI.

PROPOSITION (6.3)

$X(s)$ has the interarrival time distribution function

(a) $\qquad F(t) = \sum_{k=0}^{\infty} \mu_\alpha^m * (\mu_\alpha^n)^{*k}(t)$

and the delay time distribution function

(b) $\qquad F_T(t) = \sum_{k=0}^{\infty} (\mu_\alpha^n)^{*k}(t)\hat{\mu}_m(\alpha),$

$\qquad$ where $\mu_\alpha^m(t) = \int_0^t e^{-\alpha u}\mu_m(du).$

PROOF

Of course, the delay time is the age of the label as defined in the previous section so (b) follows by PROPOSITION (6.2). Consider now the following figure

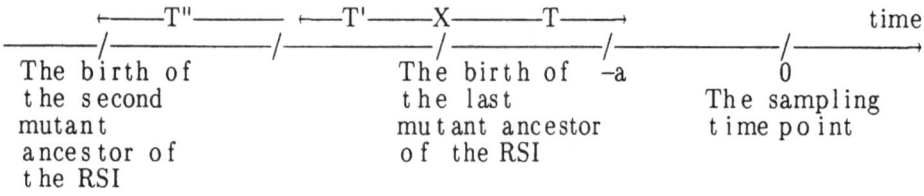

Fig. 6.2

Let T be as in the previous section, and let T' stand for the age of her mother at the birth moment of the last mutant ancestor. A suitable name for T' is the stable age at childbearing of a mutant child. Thanks to the Markov property of the sequence $(\omega_{-k}, \rho(i_k, \omega_{-k}))$, the time between the birth moment of the mother of the last mutant ancestor and that of the second mutant ancestor (T'' in the figure) will have the same distribution as T. But since T' and T are independent

$$F(t) = F_{T'} * F_T(t)$$

where

$$F_{T'}(t) = \tilde{P}(T' \leq t)$$

$$= \tilde{P}(\tau(i_k, \omega_{-k}) \leq t \mid \rho(i_k, \omega_{-k}) = 0)$$

$$= \sum_{j=1}^{\infty} \tilde{P}(\tau(i_k, \omega_{-k}) \leq t; \ i_k = j; \ \rho(i_k, \omega_{-k}) = 0)/\hat{\mu}_m(\alpha)$$

$$= E[\int_0^t e^{-\alpha u} \xi_m(du)]/\hat{\mu}_m(\alpha)$$

$$= \mu_\alpha^m(t)/\hat{\mu}_m(\alpha).$$

This gives

$$F(t) = \sum_{k=0}^{\infty} ((\mu_\alpha^n)^{*k} \hat{\mu}_m(\alpha)) * \mu_\alpha^m(t)/\hat{\mu}_m(\alpha)$$

$$= \sum_{k=0}^{\infty} \mu_\alpha^m * (\mu_\alpha^n)^{*k}(t)$$

and the statement follows.   □

Since $X(s)$ is (backwards in time) a renewal process, standard renewal theory provides us with a number of results that we can apply directly. For this purpose the mean and variance of the interarrival time distribution are needed.

PROPOSITION (6.4)

If M and V denote the mean and the variance of the interarrival time distribution in the process $X(s)$ respectively, then

$$M = \beta/\hat{\mu}_m(\alpha)$$

and

$$V = B_2/\hat{\mu}_m(\alpha) + (B_n^2 - B_m^2)/\hat{\mu}_m(\alpha)^2$$

where

$$B_2 = \int_0^{\infty} u^2 e^{-\alpha u} \mu(du),$$

B_n is as above and B_m is defined similarly.

PROOF

Since $F(t)$ is the distribution function of $T + T'$, it follows that its mean is given by

$$(B_n + B_m)/\hat{\mu}_m(\alpha) = \beta/\hat{\mu}_m(\alpha).$$

Notice that this is exactly the mean age at childbearing in the embedded label process which we calculated analytically in Chapter Four. The variance could also be derived by Laplace transformation, but we prefer the following argument: Let

$$B_{n_2} = \int_0^\infty u^2 e^{-\alpha u} \mu_n(du)$$

and

$$B_{m_2} = \int_0^\infty u^2 e^{-\alpha u} \mu_m(du),$$

then

$$\begin{aligned}
V &= \tilde{V}ar[T + T'] \\
&= \tilde{V}ar[T] + \tilde{V}ar[T'] \\
&= \tilde{E}[T^2] - \tilde{E}^2[T] + \tilde{V}ar[T'] \\
&= B_{m_2}/\hat{\mu}_m(\alpha) - B_m^2/\hat{\mu}_m(\alpha)^2 + \tilde{E}[\nu-1]\tilde{V}ar[\tau_n] + \tilde{E}[\tau_n]^2\tilde{V}ar[\nu]
\end{aligned}$$

because T' has the same distribution as what can be called the stable age at childbearing of a mutant child (τ_m) and T has the same distribution as the sum of $(\nu-1)$ i.i.d. random variables with the same distribution as what can be called the stable age at childbearing of a non-mutant child (τ_n). This gives

$$\begin{aligned}
V &= B_{m_2}/\hat{\mu}_m(\alpha) - B_m^2/\hat{\mu}_m(\alpha)^2 + \hat{\mu}_m(\alpha)(B_{n_2} - B_m^2/\hat{\mu}_m(\alpha))/\hat{\mu}_n(\alpha)\hat{\mu}_m(\alpha) \\
&\qquad + B_m^2/\hat{\mu}_n(\alpha)/\hat{\mu}_n(\alpha)\hat{\mu}_m(\alpha) \\
&= B_{m_2}/\hat{\mu}_m(\alpha) - B_m^2/\hat{\mu}_m(\alpha)^2 + B_{n_2}/\hat{\mu}_m(\alpha) - B_m^2/\hat{\mu}_m(\alpha)\hat{\mu}_m(\alpha) \\
&\qquad + B_n^2/\hat{\mu}_n(\alpha)\hat{\mu}_m(\alpha) \\
&= B_2/\hat{\mu}_m(\alpha) + (B_n^2 - B_m^2)/\hat{\mu}_m(\alpha)^2 \qquad\qquad \square
\end{aligned}$$

We can now summarize some interesting facts about $X(s)$ in the form of

THEOREM (21.4)

For large values of s,

(a) $\qquad X(s)$ can be approximated by s/M.

(b)   X(s) is approximately normally distributed with mean s/M and standard deviation $V^2 s/M^3$.

(c) $\tilde{E}[X(s+h) - X(s)] \longrightarrow h/M$, as $s \longrightarrow \infty$.

(d)   Let U_k stand for the length of the time interval separating the birth moment of the RSI and that of her kth last mutant ancestor. Then for large values of k,

U_k is, for large values of k, approximately normally distributed with mean k/M and standard deviation $V^2 k/M^3$.

PROOF

Follows from standard results in renewal theory (cf. [Ross, 1982]).   □

CHAPTER SEVEN

A MEASURE OF RELATEDNESS

Most of the results in this chapter will be formulated in terms of an RSI (cf. Chapter Five) but we will also describe what happens when the randomly sampled individual is chosen among living individuals.

Let φ denote the cardinality of the set of individuals carrying the same label as a randomly sampled individual chosen from a very old branching population. Since φ is a function of a pedigree, we can study its expectation by means of the two equivalent methods discussed in Chapter Five. Of course φ is easily and formally understood as a functional on the pedigree space $\tilde{\Omega}$ and one can apply the stable pedigree law on it directly to get its expectation $\tilde{E}[\varphi]$. Unfortunately a crucial condition in THEOREM (5.1) in [Nerman & Jagers, 1984] (cf. TH.6 in APPENDIX A) fails and we have thus to either use some other criteria that we know are satisfied directly on φ or to return to the characteristic level to actually show the convergence. A rapid investigation of the available criteria that may be used (cf. [Jagers & Nerman, 1984a]) shows that it is much more convenient to check the convergence conditions on the characteristic, χ say, "counting" φ while the limit itself is best understood in terms of the stable pedigree space. Naturally THEOREM (5.1) will guarantee that, once we have checked the conditions on the characteristic level, the formal integration $\tilde{E}[\varphi]$ with respect to $\tilde{P}$ or $\tilde{E}_\ell[\varphi]$ with respect to $\tilde{P}_\ell$ can be performed. This is done in the first section of the present chapter. In §7.2 φ's distribution will be rapidly discussed.

One of the main motivations for the study of the average size of the label of a randomly sampled individual is the possibility of using it as a measure of genetical identity within the population in genetical applications. Some authors use this average size directly as such a measure. We shall however see in §7.3 that a more appropriate measure can be defined with the average size as a main ingredient.

7.1. THE MEAN SIZE OF A LABEL

Consider the following functional of the pedigree of an RSI (the case where we sample among living individuals will be discussed later on.)

$\varphi(\tilde{\omega},a)$ = The cardinality of the set of individuals carrying the same label as the RSI.

As in Chapter Six take

$$A_n = \{(\tilde{\omega},a) \in \tilde{\Omega} \times R^+; \varphi(\tilde{\omega},a) \text{ is determined by } (p_n\tilde{\omega},a)\}$$

$$\doteq \{(\tilde{\omega},a); \text{ at least one } \rho(i_k,\omega_{-k}) \text{ is zero, } k \leq n\}$$

where p_n stands for the projection $\tilde{\Omega} \longrightarrow \tilde{\Omega}_n$. Define

$$\nu = \nu_\varphi(\tilde{\omega},a) = \inf\{n; (\tilde{\omega},a) \in A_n\}$$

(notice that ν is still the generation age of the label of the RSI as it was defined earlier and recall that $\nu = 1$ means that $\rho(i_1, \omega_{-1}) = 0$).

For individuals x and y such that x stems from y take

$$\Gamma_{y,x} = \begin{cases} 1 & \text{if x and y carry the same label} \\ \\ 0 & \text{otherwise} \end{cases}$$

Then

$$\varphi(\tilde{\omega},a) = \sum_{x \in J_\nu} \Gamma_{-\nu,x} \, 1_{R^+}(-\tilde{\sigma}_x).$$

φ can now be written in the following form

$$\varphi(\tilde{\omega},a) = \sum_{k=0}^{\infty} \sum_{x \in J_k} (1 - \rho(i_{k+1}, \omega_{-(k+1)})) \prod_{j=1}^{k} \rho(i_j, \omega_{-j})$$

$$\Gamma_{-k,x} \, 1_{R^+}(-\tilde{\sigma}_x)$$

$$= \sum_{k=0}^{\infty} \varphi_k(\tilde{\omega},a)$$

where

$$\varphi_k = 1_{\{k\}}(\nu)\varphi.$$

Let us now consider the random characteristic χ that counts φ. If we "enter" the original population at time t and associate with every individual, x, in the process the well defined part of her corresponding φ–value

$$\varphi_x(t) = \varphi(\Pi_x, t-\sigma_x) \, 1(\nu_{\varphi_x} \leq n(x))$$

where

$$\nu_{\varphi_x} = \inf\{k; \rho(x_{n(x)-k+1}, \omega_{x_{[k]}}) = 0\}$$

then clearly φ_x, for most x's, is concerned with the local behaviour of the population among x's close relatives.

Formally

$$\varphi_x(t) = \sum_{k=0}^{n(x)} 1_{\{k\}}(\nu_{\varphi_x}) \sum_{y \in I} \Gamma_{y \circ S_{x_{[k]}}} 1_{R^+}(t-\sigma_x) 1_{R^+}(t-\sigma_{x_{[k]}y})$$

where we use the following notation:

$$\Gamma_{0,x} = \Gamma_x = \prod_{j=1}^{n(x)} \rho(x_j, \omega_{x_1 \ldots x_{j-1}})$$

and generally, when $x = yz$

$$\Gamma_{y,x} = \Gamma_z \circ S_y.$$

THEOREM (7.1)

Let φ be as above and assume that the underlying branching process is well behaved in the sense of §1.5 and satisfies (C.7) of the same section. Then if $y_n(t)$ denotes the total population process associated with the ancestral process (cf. Chapter Two),

(a) $$\tilde{E}[\varphi] = \hat{\mu}_m(\alpha) \int_0^\infty \alpha e^{-\alpha t} E[y_n(t)^2] dt$$

(b) $$Z_t^\varphi / y_t \xrightarrow{\quad P \quad} \tilde{E}[\varphi], \ t \to \infty \ \text{on} \ \{y_t \to \infty\}$$

where $Z_t^\varphi = \sum_{x \in I} \varphi_x(t)$ and $\tilde{E}[\varphi]$ possibily infinite.

Assume further that there is a γ such that

(1) $\gamma < \frac{\alpha}{2}$

(2) $E[\xi_n(\gamma)] < 1$

(3) $E[\xi_n(\gamma)^2] < \infty.$

Then

(c) $\tilde{E}[\varphi] < \infty.$

PROOF

The first step of the proof consists in writing Z_t^φ as a branching process counted by a suitable random characteristic and thus in the form Z_t^χ where χ is a true characteristic. The usual resummation trick yields

$$Z_t^\varphi = \sum_{\ell=0}^\infty \sum_{x \in N^\ell} 1_{R^+}(t - \sigma_x)$$

$$\sum_{k=0}^\ell (1_{\{k\}}(\nu_{\varphi_x}) \sum_{y \in I} \Gamma_y \circ S_{x_{[k]}} 1_{R^+}(t - \sigma_{x_{[k]}y}))$$

$$= \sum_{\ell=0}^\infty \sum_{k=0}^\ell \sum_{x \in N^{\ell-k}} \sum_{z \in N^k} 1_{\{k\}}(\nu_{\varphi_x} \circ S_z) \sum_{y \in I} \Gamma_y \circ S_z$$

$$1_{R^+}(t - \sigma_z) 1_{R^+}(t - \sigma_{zy})$$

$$= \sum_{k=0}^\infty \sum_{x \in N^k} \sum_{\ell=0}^\infty 1_{\{\ell\}}(\nu_{\varphi_x} \circ S_z) \sum_{y \in I} \Gamma_y \circ S_z$$

$$1_{R^+}(t-\sigma_{zx})1_{R^+}(t-\sigma_{zy})$$

$$= \sum_{z\in I}[\sum_{\ell=0}^{\infty} \sum_{x\in N^{\ell}\{\ell\}}(\nu_{\varphi_x} \circ S_z) \sum_{y\in I}\Gamma_y \circ S_z$$

$$1_{R^+}(t-\sigma_{zy})1_{R^+}(t-\sigma_{zx})]$$

$$= \sum_{z\in I}[\sum_{i\in N}\sum_{\ell=0}^{\infty} \sum_{x\in N^{\ell}\{\ell\}}(\nu_{\varphi_x} \circ S_z) \sum_{y\in I}\Gamma_y \circ S_{zi}$$

$$1_{R^+}(t-\sigma_{ziy})1_{R^+}(t-\sigma_{zix})]$$

and this is of course of the form $\sum_{z\in I}\chi \circ S_z(t-\sigma_z)$ where $\chi \circ S_z$ is the expression within the square brackets at the right hand side of the last equality. What $\chi \circ S_z$ actually counts is the number of individuals x having zi as their last mutant ancestor for some $i \in N$. Those individuals constitute the totality of individuals carrying the label initiated by zi if this initiates any. Every such x is given a score equal to the number of individuals carrying the label in question (i.e. the y individuals in the above formulas). Now

$$Z_t^{\varphi} = Z_t^{\chi} = \sum_{z\in I}[\sum_{i\in N}(1-\rho(i,\omega_z))1_{R^+}(t-\sigma_{zi})(\sum_{x\in I}\Gamma_x \circ S_{zi}$$

$$1_{R^+}(t-\sigma_{zix}))(\sum_{y\in I}\Gamma_y \circ S_{zi}1_{R^+}(t-\sigma_{ziy})]$$

$$= \sum_{z\in I}[\sum_{i\in N}(1-\rho(i,\omega_z))1_{R^+}(t-\sigma_{zi})(\sum_{x\in I}\Gamma_x \circ S_{zi}1_{R^+}(t-\sigma_{zix})^2]$$

$$= \sum_{z\in I}[\sum_{i\in N}(1-\rho(i,\omega_z))y_n^2(t-\sigma_{zi}) \circ S_{zi}]$$

$$= \sum_{z\in I}\chi_z(t-\sigma_z)$$

where

$$\chi(t) = \sum_{i\in N}(1-\rho(i,\omega_0))y_n^2(t-\tau(i,\omega_0)) \circ S_i.$$

By THEOREM (5.1), we can identify $\tilde{E}[\varphi]$ with

$$E[\hat{\chi}(\alpha)] = \hat{\mu}_m(\alpha)\int_0^{\infty} \alpha e^{-\alpha t}E[y_n^2(t)]dt$$

and this concludes (a).

In order to show (b) notice that the finiteness of $\tilde{E}[\varphi]$ is equivalent to the direct Riemann integrability (d.r.i) of $e^{-\alpha t}E[\chi(t)]$. This is indeed due to the fact that $E[y_n^2 * \xi_m(t)]$ is monotonic which implies that $e^{-\alpha t}E[\chi(t)]$ is d.r.i if

$$\int_0^{\infty}e^{-\alpha t}\int_0^{\infty}E[y_n^2(t-u)\,\xi_m(du)]du < \infty$$

or equivalently if

$$\hat{\mu}_m(\alpha) \int_0^\infty e^{-\alpha t} E[y_n^2(t)]dt < \infty$$

which is the same as

$$\tilde{E}[\varphi] < \infty.$$

It is thus seen that (b) will follow if $\tilde{E}[\varphi]$ is finite. If on the other hand $\tilde{E}[\varphi]$ is not finite the convergence will still hold as can be seen using the following rapid truncation agument: Consider

$$\varphi_C = \varphi 1_{[0,C]} \uparrow \varphi \quad \text{as } C \uparrow \infty.$$

Then clearly

$$Z_t^{\varphi_C}/y_t \le Z_t^{\varphi}/y_t$$

and since φ_C is easily seen to "correspond" to a random characteristic that satisfies the requirements of TH.3 in APPENDIX A, we can conclude that

$$Z_t^{\varphi_C}/y_t \xrightarrow{P} \tilde{E}[\varphi 1_{[0,C]}] \quad \text{on } \{y_t \to \infty\}, t \to \infty.$$

We can now achieve the desired convergence by letting $C \uparrow \infty$. The theorem will now follow if we can show that the requirement

$$\int_0^\infty e^{-\alpha t} E[y_n^2(t)] < \infty$$

is satisfied under assumptions (1)–(3). This is rather technical and is therefore done in a separate lemma. □

LEMMA (7.2)

If there is a γ satisfying

1) $\qquad \gamma < \frac{\alpha}{2}$

2) $\qquad E[\xi_n(\gamma)] < 1$

3) $\qquad E[\xi_n(\gamma)^2] < \infty,$

Then

$$\int_0^\infty e^{-\alpha t} E[y_n^2(t)]dt < \infty.$$

PROOF

Notice to begin with that the assertion will follow if we can show that

$$\lim_{u \to \infty} \sup e^{-2\gamma u} E[y_n^2(u)] < \infty.$$

But since

$$E[y_n^2(u)] = (E[y_n(u)])^2 + \text{Var}[y_n(u)],$$

the problem reduces to the study of whether

$$\lim_{u \to \infty} \sup e^{-2\gamma u} \mathrm{Var}[y_n(u)] < \infty.$$

Just notice that the assumption $\hat{\mu}_n(\gamma) < 1$ yields that

$$\int_0^\infty e^{-\gamma u} E[y_n(u)]du = 1/(1-\hat{\mu}_n(\gamma)) < \infty,$$

and thereby the monotonicity of $E[y_n(u)]$ ensures that

$$\lim_{u \to \infty} e^{-\gamma u} E[y_n(u)] = 0.$$

In order to show the stronger than needed

$$\lim_{u \to \infty} e^{-2\gamma u} \mathrm{Var}[y_n(u)] = 0,$$

we can follow [Jagers & Nerman, 1984a] and define

$$h_n(t) = \mathrm{Var}[\psi_0(t) + \int_0^t E[y_n(t-u)]\xi_{n,0}(du)]$$

where $\psi = 1_{[0,\infty)}$ is the (individual) random characteristic "counting" the general

branching process $y_n(t)$. Recalling that $\nu_n(t) = \sum_{k=0}^\infty \mu_n^{*k}(t)$, we can now write

$$e^{-2\gamma t} \mathrm{Var}[y_n(t)] = \int_0^t e^{-2\gamma(t-u)} h_n(t-u) e^{-2\gamma u} \nu_n(du).$$

Clearly, $e^{-2\gamma t} h_n(t)$ equals

$$\mathrm{Var}[\int_0^t e^{-\gamma t} E[y_n(t-u)]\xi_n(du)]$$

$$= \mathrm{Var}[\int_0^\infty 1_{[0,t]}(u) \, e^{-\gamma(t-u)} E[y_n(t-u)] e^{-\gamma u} \xi_n(du)].$$

Since

$$\sup_u e^{-\gamma u} E[y_n(u)] < \infty$$

and

$$E[\xi_n(\gamma)^2] < \infty,$$

it is seen that

$$\sup_t e^{-2\gamma t} h_n(t) < \infty$$

and further dominated convergence yields

$$\lim_{t \to \infty} e^{-2\gamma t} h_n(t) = \lim_{t \to \infty} \mathrm{Var}[\int_0^t e^{-\gamma t} E[y_n(t-u)]\xi_n(du)] = 0.$$

Another application of dominated convergence, and we get

$$\lim_{t \to \infty} e^{-2\gamma t} \mathrm{Var}[y_n(t)] = 0$$

because

$$\int_0^\infty e^{-2\gamma u}\nu_n(du) < 1 \,/\, (1 - \hat{\mu}_n(\gamma)) < \infty. \qquad \square$$

A more direct proof of the fact that

$$\tilde{E}[\varphi] = \hat{\mu}_m(\alpha) \int_0^\infty \alpha e^{-\alpha t} E[y_n^2(t)]dt$$

can be obtained by applying the stable pedigree measure directly using LEMMA (5.2):

$$\tilde{E}[\varphi] = \tilde{E}[\sum_{k=0}^\infty \sum_{x \in J_k} (1 - \rho(i_{k+1}, \omega_{-(k+1)})$$

$$\prod_{\ell=1}^k \rho(i_\ell, \omega_{-\ell}) \, \Gamma_{-k,x} \, 1_{R^+}(-\sigma_x)]$$

$$= \sum_{k=0}^\infty \tilde{E}[y_{n,-k}(a + \tau(i_1, \omega_{-1}) + \ldots + \tau(i_k, \omega_{-k}))$$

$$\prod_{\ell=1}^k \rho(i_\ell, \omega_{-\ell})(1 - \rho(i_{k+1}, \omega_{-(k+1)}))].$$

Since each expression of the form

$$y_{n,-k}(a + \tau(i_1, \omega_{-1}) + \ldots + \tau(i_k, \omega_{-k}))$$

$$\prod_{\ell=1}^k \rho(i_\ell, \omega_{-\ell})(1 - \rho(i_{k+1}, \omega_{-(k+1)}))$$

is k–determined LEMMA (5.2) implies that

$$\tilde{E}[\varphi]$$

$$= \hat{\mu}_m(\alpha) \sum_{k=0}^\infty \sum_{x \in N^k} E[e^{-\alpha\sigma_x} \int_0^\infty \alpha e^{-\alpha t} y_n(t+\sigma_x)dt$$

$$\prod_{\ell=1}^k \rho(x_\ell, \omega_{x_1 \ldots x_{\ell-1}})]$$

$$= \hat{\mu}_m(\alpha) \, E[\int_0^\infty e^{-\alpha u} \int_0^\infty \alpha e^{-\alpha t} y_n(t+u)dt \, y_n(du)]$$

$$= \hat{\mu}_m(\alpha) \, E[\int_0^\infty \int_0^\infty \alpha e^{-\alpha(t+u)} y_n(t+u)dt \, y_n(du)]$$

$$= \hat{\mu}_m(\alpha) \, E[\int_0^\infty \int_u^\infty \alpha e^{-\alpha v} y_n(v)dv \, y_n(du)]$$

$$= \hat{\mu}_m(\alpha) \, E[\int_0^\infty \int_0^v y_n(du) \alpha e^{-\alpha v} y_n(v)dv]$$

$$= \hat{\mu}_m(\alpha) \; E[\int_0^\infty \alpha e^{-\alpha v} \, y_n(v)^2 dv]$$

$$= \hat{\mu}_m(\alpha) \int_0^\infty \alpha e^{-\alpha v} E[y_n^2(v)]dv,$$

as expected.

In order to get the (stable) mean number of living individuals, carrying the same label as a random individual chosen among all those alive at some late sampling time point, it turns out that the proof of THEOREM (7.1) goes through with the slight modification that the x:s and the y:s in the formulas there have to be alive. This is achieved by introducing two convenient indicator functions everywhere (cf. the remarks at the end of Chapter Five), resulting in formulas like

$$\sum_{z \in I} \sum_{i \in N} (1-\rho(i,\omega_z)) \left[\sum_{x \in I} \Gamma_x \circ S_{zi} 1_{[0,\lambda_{zix}]}(t-\sigma_{zix}) \right]$$

$$\left[\sum_{y \in I} \Gamma_y \circ S_{zi} 1_{[0,\lambda_{ziy}]}(t-\sigma_{ziy}) \right] \cdot$$

$$= \sum_{z \in I} \left[\sum_{i \in N} (1-\rho(i,\omega_z))(Z^n_{t-\sigma_{zi}} \circ S_{zi})^2 \right]$$

where Z^n_t is as in Chapter Two. The final formula is

$$\tilde{E}_\ell[\varphi] = \hat{\mu}_m(\alpha) \int_0^\infty \alpha e^{-\alpha t} E[(Z^n_t)^2]dt \; / \; (1 - \hat{L}(\alpha)),$$

where L stands for the life span distribution function. Similarly, if we are interested in the mean number of living individuals carrying the same label as an RSI, we get

$$\hat{\mu}_m(\alpha) \int_0^\infty \alpha e^{-\alpha t} E[y_n(t)Z^n_t]dt.$$

Let us now consider an explicit example. Think of the case where the underlying process is a Bellman-Harris branching process with exponentially (δ) distributed life spans and reproduction p.g.f. f. This is often referred to as the splitting Markovian case. Let us further assume independent (θ) mutations. Then the p.g.f. describing the reproduction of non-mutants will be

$$g(s) = f(\theta + (1-\theta)s).$$

Notice that in this case

$$\alpha = \delta(m-1), \quad m = \sum_{k=1}^\infty k p_k$$

and

$$\alpha_n = \alpha - \delta m \theta.$$

if we now define

$$h(s) = \delta(g(s) - s)$$

then we can use a classical result in [Athreya & Ney, 1972] stating that if

$$\sum_{k=1}^{\infty} k^2 p_k < \infty$$

then

$$m_2^{\mathbf{n}}(t) = E[(Z_t^{\mathbf{n}})^2]$$

will be finite for all t, and can be shown to satisfy the backward equation

$$\frac{d}{dt} m_2^{\mathbf{n}}(t) = h''(1) e^{2\alpha_n t} + \delta m_2^{\mathbf{n}}(t)$$

with the boundary condition $m_2^{\mathbf{n}}(0) = 1$.

The solution of the above equation turns out to be $m_2^{\mathbf{n}}(t) = h''(1)t$ if $\alpha_n = 0$

and $m_2^{\mathbf{n}}(t) = h''(1) \alpha_n^{-1}(e^{2\alpha_n t} - e^{\alpha_n t})$ otherwise. Obviously this can be used to obtain $\tilde{E}_{\ell}[\varphi]$ for this particular process.

7.2. THE DISTRIBUTION OF THE SIZE OF A LABEL

Consider a well behaved branching process and let φ be defined as in the preceding section. In what follows we derive an expression for the distribution of the size of the label of an RSI.

In the same fashion as in the preceding section it is possible to formulate similar results for the case where the sampling category is restricted to those alive at the sampling time point.

THEOREM (7.3)

$$\tilde{P}(\varphi=j) = \hat{\mu}_{\mathbf{m}}(\alpha) \int_0^{\infty} \alpha e^{-\alpha t} j P(y_{\mathbf{n}}(t) = j) dt$$

PROOF

Take $\psi_j = 1_{[\varphi=j]}$. Since TH.6 in APPENDIX A applies for indicators, we can use the stable pedigree measure directly and get appropriate convergence towards

$$\tilde{P}(\varphi=j) = \tilde{E}[\psi_j]$$

$$= \tilde{E}[\sum_{k=0}^{\infty} 1_{\{j\}}(y_{\mathbf{n},-k}(a+...+\tau(i_k,\omega_{-k})) \prod_{\ell=1}^{k} \rho(i_{\ell},\omega_{-\ell})$$

$$(1 - \rho(i_{k+1},\omega_{-(k+1)}))]$$

as was seen at the end of the preceding section. Applying LEMMA (5.2) will now give

$$\tilde{P}(\varphi=j) = \hat{\mu}_{\mathbf{m}}(\alpha) \sum_{k=0}^{\infty} \sum_{x \in N^k} E[e^{-\alpha \sigma_x} \int_0^{\infty} \alpha e^{-\alpha t} 1_{\{j\}}(y_{\mathbf{n}}(t+\sigma_x)) dt$$

$$\prod_{\ell=1}^{k} \rho(x_{\ell}, \omega_{x_1...x_{\ell}})].$$

Similar calculations as in the study of $\tilde{E}[\varphi]$ can be done to yield

$$\tilde{P}(\varphi=j)$$

$$= \hat{\mu}_m(\alpha) \; E[\int_0^\infty \int_0^\infty \alpha e^{-\alpha t} e^{-\alpha u} \; 1_{\{j\}}(y_n(t+u)) dt \; y_n(du)]$$

$$= \hat{\mu}_m(\alpha) E[\int_0^\infty \alpha e^{-\alpha t} 1_{\{j\}}(y_n(t)) y_n(t) dt]$$

$$= \hat{\mu}_m(\alpha) E[\int_0^\infty \alpha e^{-\alpha t} \; j \; 1_{\{j\}}(y_n(t)) dt]$$

$$= \hat{\mu}_m(\alpha) \int_0^\infty \alpha e^{-\alpha t} j P(y_n(t)=j) dt. \qquad \qquad \square$$

That $p_j = \hat{\mu}_m(\alpha) \int_0^\infty \alpha e^{-\alpha t} j P(y_n(t)=j) dt$, $j=0,1,2,...$ define a probability measure is seen

by

$$\sum_{j=0}^\infty p_j = \hat{\mu}_m(\alpha) \int_0^\infty \alpha e^{-\alpha t} \sum_{j=0}^\infty j \, P(y_n(t)=j) dt$$

$$= \hat{\mu}_m(\alpha) \int_0^\infty \alpha e^{-\alpha t} E[y_n(t)] dt$$

$$= \hat{\mu}_m(\alpha) \sum_{k=0}^\infty \hat{\mu}_n(\alpha)^k$$

$$= \hat{\mu}_m(\alpha)/(1 - \hat{\mu}_n(\alpha))$$

$$= 1.$$

We can now derive a formula giving the different moments of the size of the label of the RSI (providing those moments do exist). Formally

$$\tilde{E}[\varphi^k] = \sum_{j=0}^\infty j^k p_j$$

$$= \hat{\mu}_m(\alpha) \int_0^\infty \alpha e^{-\alpha t} \sum_{j=0}^\infty j^{k+1} P(y_n(t)=j) dt$$

$$= \hat{\mu}_m(\alpha) \int_0^\infty \alpha e^{-\alpha t} E[y_n^{k+1}(t)] dt,$$

which, when $k=1$, agrees with what we already know.

7.3. THE PROBABILITY OF IDENTITY BY DESCENT

In the introduction to the present chapter, it was mentioned that the average size of the label of a randomly sampled individual can be used to define a measure of the overall variation over labels in the population. In order to make this idea precise we recall that the usual measure of relatedness (or equivalently unrelatedness or variation) is the probability of identity by descent. In our case this can be defined as the probability that two randomly sampled individuals chosen among those living at the sampling time point carry the same label. In this case we say that they are identical by descent. Clearly this occurs precisely if they stem from a common ancestor without intervening mutations. In what follows we describe a version of this measure. Let 0_1 and 0_2 denote the two sampled individuals (assumed to be sampled independently and with replacement). Let further $F_{0_1}(t)$ stand for the family of individuals carrying the same label as 0_1 and define P_t as the conditional probability of the event that 0_1 and 0_2 are identical by descent, given the branching population evolution. Then P_t can be approximated by a population size determined expression according to the following theorem where Z_t stands for the number of living individuals at time t.

THEOREM (7.4)

Under the conditions of THEOREM (7.1), and for large values of t,

$$P_t \sim \tilde{E}_\ell[\varphi] / Z_t$$

in the sense of convergence in probability of the ratio towards 1, on the set of non–extinction.

PROOF

By definition

$$P_t = \text{Prob}(0_2 \in F_{0_1}(t))$$

where Prob is used to denote that probability is induced by the sampling mechanism and not the population law.

$$= \sum_{x \in I} \text{Prob}(0_2 \in F_{0_1}(t)|0_1 = x)\text{Prob}(0_1 = x)$$

$$= \sum_{x \in I} \frac{\varphi(\Pi_x, \, t-\sigma_x)}{Z_t} \cdot \frac{1}{Z_t}$$

By the discussion following THEOREM (7.1)

$$\frac{1}{Z_t} \sum_{x \in I} \varphi(\Pi_x, t-\sigma_x) \longrightarrow \tilde{E}_\ell[\varphi]$$

in various senses on $\{y_t \to \infty\}$. □

REMARKS

− A slightly different measure of (label) relatedness will be given in the next chapter.

− The situation above can be abstracted to an urn–ball situation which can be used to heuristically understand the approximation above.

− In Chapter Eleven the natural generalization of the problem treated in the present section to sampling of k individuals is briefly discussed.

CHAPTER EIGHT

INFINITE SITES LABELS

In this chapter the idea of a label will be made more precise. As has already been pointed out, models with infinitely many labels have their motivation in the fine structure of the gene. This fine structure seems to suggest that a label may be viewed as a very long sequence of specific positions (sites). Each such position can be occupied by some letter chosen from a four–letter alphabet (the four nucleotides A,C,T and G). Models adopting this special point of view go back to [Kimura, 1969] and [Kimura, 1971] and are often named infinite sites models.

For simplicity, we assume that a label can be seen as a long sequence of sites as described above and work under the further assumption that every mutation affects only one site (this assumption can be argued for as in [Kimura, 1971]).

To sum up, we consider a model where every mutant individual carries an entirely new label differing from the parental one with respect to exactly one site, and from her potential mutant sisters with respect to exactly two sites. This means that we disregard recurrent mutations. Now as before what really interests us is the variation over labels within the population counted in several senses and not the labels themselves. It is also to be noticed that since this new situation is essentially the one discussed in the previous chapters we can still use our usual terminology.

In this new framework we can define a natural measure of genetic distance between pairs of individuals through

$$d(x,y):= \text{ the number of sites with respect to which}$$
$$x \text{ and } y \text{ differ.}$$

The present chapter will be devoted to the study of the family of (both dead and living) individuals at some given genetic distance, d, from an RSI (recall the convention defining an RSI from Chapter Five) in an old branching population. For convenience this set of individuals will be referred to as the F(d) family of the RSI. Obviously, once we have studied the F(d) family it will be easy to make statements about

$$F'(d) = \text{ the family of individuals differing from the}$$
$$\text{RSI with respect to at most } d \text{ sites.}$$

Also in the present case it is possible to allow sampling in the restricted set of individuals containing those still alive at the sampling time point. As before this is achieved either by conditioning in the stable pedigree measure or by adding indicator functions $1(\lambda_x > t - \sigma_x)$ in the expressions for processes counted by random characteristics.

8.1.   THE MEAN SIZE

In Chapter Seven some results were obtained concerning the size of the label carried by an RSI in an old branching population. An obvious manner of generalizing this quantity with biological relevence is to consider the size of the $F(d)$ family of an RSI. Formally this can be expressed as

$$F(d) = \{x \in J; d(x,O) = d; \tilde{\sigma}_x \leq 0\}.$$

Then the subject of study of the previous chapter reduces to that of the cardinality of $F(0)$.

It turns out to be convenient to begin the investigation by considering, for $k = 0,1,..., i = 0,1,...$

$F(k,i) =$ The family of individuals born before the sampling
time point and stemming from the $(k+1)$th last
mutant ancestor of the RSI and differing from this
with respect to exactly i sites exception made
for the progeny of her kth last mutant
ancestor.

With

$-k' =$ The kth last mutant ancestor of the RSI,

we can write

$$F(k,i) = \{x \in \bigcup_{j=k'+1}^{(k+1)'} \{-j\} \times I; d(x,-(k+1)') = i \ \& \ \tilde{\sigma}_x \leq 0\}.$$

The cardinalities of the sets $F(d)$, $F'(d)$ and $F(k,i)$ will be denoted $f(d)$, $f'(d)$ and $f(k,i)$ respectively. It is not very hard to realize that

$$f(d) = \sum_{k=0}^{d} f(k, d-k)$$

and

$$f'(d) = \sum_{i=0}^{d} f(i).$$

This motivates the definition of $F(k,i)$ since clearly it is the key to both $F(d)$ and $F'(d)$.

Since $f(k,i)$ is functional of the pedigree of an RSI we study it (and the same could of course be said of $f(d)$ and $f'(d)$) using the two equivalent fashions discussed in Chapter Five (cf. also Chapter Seven). We prefer here not to work out the details and refer the reader to those chapters.

There are two main differences between the situation we consider in this chapter and that of the previous one. The first one is the nature of the backward stopping time, and the second one is the way in which we count the individuals.

The following notation will be used

When x stems from y take

$$\Sigma_{y,x} = d(x,y)$$

and

$$\Sigma_x = \Sigma_{0,x}.$$

Then

$$\Sigma_x := \sum_{\ell=1}^{n(x)} (1 - \rho(x_\ell, \omega_{x_1,...,x_{\ell-1}}))$$

and generally when $x = yz$,

$$\Sigma_{y,x} = \Sigma_z \circ S_y.$$

We will also take

$$\nu_k^x := \min\{j \geq 1; \sum_{\ell=n(x)-j+1}^{n(x)} (1 - \rho(x_\ell, \omega_{x_1...x_{\ell-1}})) = k\}.$$

Let us now associate with each individual, x, in the process the contribution $f^{k,i}(\Pi_x, t - \sigma_x)$ where $f^{k,i}$ is just another notation for $f(k,i)$. Summation over all individuals born before t yields the process

$$Z_t^{k,i} = \sum_{x \in I} f^{k,i}(\Pi_x, t-\sigma_x) 1(\nu_k^x \leq n(x)).$$

THEOREM (8.1)

Let

$$f^{k,i}(\tilde{\omega},a) =$$
$$(k+1)'$$
$$= \sum_{j=k'+1} \sum_{x \in \{-j\} \times I} 1_{\{i\}}(\Sigma_{-j,x}) 1_{R^+}(-\tilde{\sigma}_x)$$

and assume that the underlying process is well behaved in the sense of Chapter One and satisfies (C.7) in that chapter. Then

(a) $\tilde{E}[f(k,i)] = \hat{\mu}_m(\alpha)\int_0^\infty \alpha e^{-\alpha t} E[y^i(t)y^k(t)]dt -$

$$\hat{\mu}_m(\alpha)\int_0^\infty \alpha e^{-\alpha t} E[y^{i-1}(t)y^{k-1}(t)]dt,$$

$i = 0,1,...$ and $k=0,1,...$

where $y^i(t)$ is the process counting individuals x born before t and satisfying $d(0,x) = i$ (here 0 refers to the original ancestor of the population), and $y^{-1}(t) = 0$.

(b) $Z_t^{k,i}/y_t \xrightarrow{P} \tilde{E}[f(k,i)]$ as $t \to \infty$ on $\{y_t \to \infty\}$

where $Z_t^{k,i}$ is as above and $\tilde{E}[f(k,i)]$ possibly infinite.

Assume further that there exists some γ such that

(i) $\gamma < \frac{\alpha}{2}$

(ii) $E[\hat{\xi}_n(\gamma)] < 1$

(iii) $E[\hat{\xi}(\gamma)^2] < \infty$

Then

c) $\tilde{E}[f(k,i)] < \infty$

PROOF

Ignoring individuals with exactly k mutant ancestors, $Z_t^{k,i}$ can be approximated by the following sum to be referred to as $Z_t^{k,i}$

$$\sum_{x\in I} \left(\sum_{y\in I} 1_{\{i\}}(\Sigma_y \circ S_{x_{[\nu_{k+1}^x -1]}})1_{R^+}(t-\sigma_{x_{[\nu_{k+1}^x -1]}}y) \right.$$

$$\left. - \sum_{u\in I} 1_{\{i-1\}}(\Sigma_u \circ S_{x_{[\nu_k^x -1]}})1_{R^+}(t-\sigma_{x_{[\nu_k^x -1]}}y) \right).$$

Since the two terms of the above expression are similar we choose to study the first one. This term is transformed into a branching process counted by a random characteristic by means of the following rewritings. First write it as

$$\sum_{x\in I} \sum_{h=0}^{n(x)} 1_{\{h\}}(\nu_{k+1}^x) \sum_{y\in I} 1_{\{i\}}(\Sigma_y \circ S_{x_{[h-1]}})1_{R^+}(t-\sigma_{x_{[h-1]}}y)$$

and take $x = x'jx''$, $x' \in N^{n(x)-h}$, $j\in N$, $x'' \in N^{h-1}$ to get

$$= \sum_{n=0}^{\infty} \sum_{h=0}^{n} \sum_{x'\in N^{n-h}} \sum_{j\in N} \sum_{x''\in N^{h-1}} 1_{\{h\}}(\nu_{k+1}^{x'jx''})$$

$$\sum_{y\in I} 1_{\{i\}}(\Sigma_y \circ S_{x'j})1_{R^+}(t-\sigma_{x'jy})1_{R^+}(t-\sigma_{x'jx''})$$

$$= \sum_{n=0}^{\infty} \sum_{h=0}^{\infty} \sum_{x'\in N^{n-h}} \sum_{j\in N} (1-\rho(j,\omega_{x'}))$$

$$\sum_{x''\in N^{h-1}} 1_{\{k\}}(\Sigma_{x''} \circ S_{x'j})1_{\{i\}}(\Sigma_y \circ S_{x'j})1_{R^+}(t-\sigma_{x'jy})$$

$$1_{R^+}(t-\sigma_{x'jx''}).$$

$$= \sum_{n=0}^{\infty} \sum_{x'\in N^n} \sum_{j\in N} \sum_{h=0}^{\infty} \sum_{x''\in N^h} (1-\rho(j,\omega_{x'}))$$

$$1_{\{k\}}(\Sigma_{x''} \circ S_{x'j})1_{R^+}(t-\sigma_{x'jx''}) \sum_{y\in I} 1_{\{i\}}(\Sigma_y \circ S_{x'j})$$

$$1_{R^+}(t-\sigma_{x'y})$$

$$= \sum_{x\in I} [\sum_{j\in N} (1-\rho(j,\omega_x))y^i(t-\sigma_{xj}) \circ S_{xj}y^k(t-\sigma_{xj}) \circ S_{xj}],$$

where $y^i(\cdot)$ is as in the statement of the theorem. Notice that the last expression is of the form $Z_t^{\chi'}$. Moreover

$$E[\hat{\chi}'(\alpha)] = E[\int_0^\infty \alpha e^{-\alpha t} \sum_{j \in N} (1 - \rho(j,\omega_0)) y^i(t-\tau(j)) \circ S_j$$
$$y^k(t-\tau(j)) \circ S_j dt].$$

Using the basic decomposition yields

$$E[\hat{\chi}'(\alpha)] = E[\int_0^\infty \alpha e^{-\alpha t} \sum_{j \in N} (1 - \rho(j,\omega_0)) y^i(t-\tau(j)) \circ S_j$$
$$y^k(t-\tau(j)) \circ S_j dt | \mathscr{b}_0]]$$

which, by the branching property, equals (with the sum $\sum_{j \in N}$ written as an integral)

$$\int_0^\infty \alpha e^{-\alpha t} \int_0^t E[y^i(t-u)y^k(t-u)]E[\xi_m(du)]dt = \hat{\mu}_m(\alpha) \int_0^\infty \alpha e^{-\alpha t} E[y^i(t)y^k(t)]dt.$$

To sum up we have proved that $Z_t^{k,i}$ equals

$$= \sum_{x \in I} [\sum_{j \in N} (1-\rho(j,\omega_x))(y^i(t-\sigma_{xj}) \circ S_{xj} y^k(t-\sigma_{xj}) \circ S_{xj}$$
$$- y^{i-1}(t-\sigma_{xj}) \circ S_{xj} y^{k-1}(t-\sigma_{xj}) \circ S_{xj})] = Z_t^\chi$$

where χ is the expression within the square bracket. We have also proved that the limit of Z_t^χ/y_t has the form

$$E[\hat{\chi}(\alpha)] = \hat{\mu}_m(\alpha) \int_0^\infty \alpha e^{-\alpha t} E[y^i(t)y^k(t) - y^{i-1}(t)y^{k-1}(t)]dt.$$

It is now not hard to see that the above is also the limit of $Z_t^{k,i}/y_t$. Now,

$$E[\hat{\chi}(\alpha)] = \hat{\mu}_m(\alpha)\int_0^\infty \alpha e^{-\alpha t} E[y^i(t)y^k(t)-y^{i-1}(t)y^{k-1}(t)]dt$$

combined with THEOREM (5.1) gives (a). In order to obtain (b) we refer the reader to the discussion in the proof of THEOREM (7.1) together with TH.3 in APPENDIX A. As was also discussed in the proof of THEOREM (7.1) the direct Riemann integrability of $e^{-\alpha t}E[\hat{\chi}(\alpha)]$ and the finiteness of $\check{E}[f(k,i)]$ are obtained once we have proved that

$$\int_0^\infty \alpha e^{-\alpha t} E[y^i(t)y^k(t) - y^{i-1}(t)y^{k-1}(t)]dt$$

is finite. This turns out (just as was the case with THEOREM (7.1)) to be rather technical, and we prefer to do it in a separate lemma.   □

LEMMA (8.2)

If there exists a γ such that

1) $\gamma < \dfrac{\alpha}{2}$

2) $E[\hat{\xi}_n(\gamma)] < 1$

3) $E[\hat{\xi}(\gamma)^2] < \infty,$

then $\int\limits_0^\infty \alpha e^{-\alpha t} E[y^i(t)y^k(t)]dt < \infty$

PROOF

By Schwartz's inequality for expectations, with $\alpha = 2\gamma + \epsilon$

$$\int\limits_0^\infty e^{-\alpha t} E[y^i(t)y^k(t)]dt \le \int\limits_0^\infty \sqrt{e^{-\alpha t}E[y^i(t)^2]}\;\sqrt{e^{-\alpha t}E[y^k(t)^2]}dt.$$

$$= \int\limits_0^\infty e^{-\epsilon t}\sqrt{e^{-2\gamma t}E[y^i(t)^2]}\;\sqrt{e^{-2\gamma \tau}E[y^k(t)^2]}dt.$$

Since $\int\limits_0^\infty e^{-\epsilon t}dt < \infty,$ we can apply Schwartz inequality a second time and conclude that

the studied integral is finite if

$$\int\limits_0^\infty e^{-\epsilon t}e^{-2\gamma t}E[y^i(t)^2]dt$$

is finite for every fixed i.

The lemma will certainly follow if

$$\sup_t e^{-2\gamma t}E[y^i(t)^2] < \infty,$$

which is true if

(8.1.1) $\sup\limits_t e^{-2\gamma t}\mathrm{Var}[y^i(t)] < \infty$

and

(8.1.2) $\sup\limits_t e^{-\gamma t}E[y^i(t)] < \infty.$

Notice that we can write $y^i(t)$ in the form

$$\sum_{x \in N^i} \Delta_x(t-\sigma_x)$$

where $x \in N^i$ counts all individuals (labels) in the ith generation of the embedded label process and Δ_x stands for the size of x seen as a label. This implies that (8.1.2) will be proved if

$$\int\limits_0^\infty e^{-\gamma t}E[\xi_i'^{(i)}]*E[y_n](t)dt < \infty$$

where

$$\xi'^{(i)}(t) = \text{the number of individuals in the ith generation}$$
$$\text{of the embedded label process.}$$

But this is certainly true as seen from (recall that we have agreed in Chapter Four to let primed symbols refer to the embedded label process)

$$E[\hat{\xi}'^{(i)}(\gamma)] = E[\hat{\xi}'(\gamma)]^i$$

$$= (\frac{\hat{\mu}_m(\gamma)}{1-\hat{\mu}_n(\gamma)})^i < \infty$$

and

$$E[\hat{y}_n(\gamma)] = \frac{1}{1-\hat{\mu}_n(\gamma)} < \infty$$

by the condition (2) in the statement of the lemma.

We still have to show that

$$e^{-\gamma t}Var[y^i(t)] < \infty.$$

But we can write

$$e^{-\gamma t}Var[y^i(t)] = e^{-\gamma t} \, Var \, [\, \underset{x \in N^i}{\Sigma} \, \Delta_x(t-\sigma_x)]$$

and we can now use the variance decomposition and get

$$e^{-2\gamma t}Var[y^i(t)] = e^{-2\gamma t}Var[E[\underset{x \in N^i}{\Sigma} \, \Delta_x(t-\sigma_x)|\mathscr{A}_{i-1}]] +$$

$$e^{-2\gamma t}E[Var[\underset{x \in N^i}{\Sigma} \, \Delta_x(t-\sigma_x)|\mathscr{A}_{i-1}]].$$

Define $H(t)$ and $K(t)$ to be the two terms at the right hand side of the above equality, then that

$$K(t) = e^{-2\gamma t}E[\int_0^t Var[y_n(t-u)]\xi'^{(i)}(du)dt] < \infty$$

follows from the discussion in the proof of THEOREM (7.1) and the requirement $E[\hat{\xi}_n(\gamma)]$

$< 1.$

The finiteness of $H(t)$ is seen from

$$H(t) = e^{-2\gamma t}Var\left[E[\underset{x \in N^i}{\Sigma} \, \Delta_x(t-\sigma_x)|\mathscr{A}_{i-1}]\right]$$

$$= e^{-2\gamma t}Var[\int_0^t E[y_n(t-u)]\xi'^{(i)}(du)]$$

$$= Var[e^{-\gamma t} \int_0^t \nu_n(t-u)\xi'^{(i)}(du)]$$

$$= Var[\int_0^\infty e^{-\gamma(t-u)}1_{R^+}(t-u) \, \nu_n(t-u)e^{-\gamma u}\xi'^{(i)}(du)].$$

But this will be finite if

$$Var[\hat{\xi}'^{(i)}(\gamma)] < \infty$$

and thereby if $E[\hat{\xi}'(\gamma)] < \infty$ and $Var[\hat{\xi}'(\gamma)] < \infty$ as is shown in PROPOSITION (8.3).

By the proof of (iii) in theorem (4.1)

$$Var[\hat{\xi}'(\gamma)] = \hat{\mu}_n(2\gamma)Var[\hat{\xi}'(\gamma)] +$$

$$\text{Var}[\hat{\xi}_m(\gamma) + \hat{\xi}_n(\gamma) \, \frac{\hat{\mu}_m(\gamma)}{1 - \hat{\mu}_n(\gamma)}].$$

We see now that

$$\text{Var}[\hat{\xi}'(\gamma)] = \frac{1}{1 - \hat{\mu}_n(2\gamma)} \ \text{Var}[\hat{\xi}_m(\gamma) + \hat{\xi}_n(\gamma) \, \frac{\hat{\mu}_m(\gamma)}{1 - \hat{\mu}_n(\gamma)}]$$

$$\leq \frac{1}{1 - \hat{\mu}_n(2\gamma)} \ \text{E}[(\hat{\xi}_m(\gamma) + \hat{\xi}_n(\gamma) \, \frac{\hat{\mu}_m(\gamma)}{1 - \hat{\mu}_n(\gamma)})^2]$$

$$\leq \frac{1}{1 - \hat{\mu}_n(2\gamma)} \ \text{E}[(\hat{\xi}_m(\gamma) \, \frac{\hat{\mu}_m(\gamma)}{1 - \hat{\mu}_n(\gamma)} + \hat{\xi}_m(\gamma) \, \frac{\hat{\mu}_m(\gamma)}{1 - \hat{\mu}_n(\gamma)})^2]$$

(since $\gamma < \alpha$, $\hat{\mu}_n(\gamma) + \hat{\mu}_n(\gamma) = \hat{\mu}(\gamma) \geq \hat{\mu}(\alpha) = 1$ which gives $\dfrac{\hat{\mu}_m(\gamma)}{1 - \hat{\mu}_n(\gamma)} \geq 1$)

$$\leq \frac{1}{1 - \hat{\mu}_n(2\gamma)} \ \frac{\hat{\mu}_m(\gamma)^2}{(1 - \hat{\mu}_n(\gamma))^2} \ \text{E}[\hat{\xi}(\gamma)^2]$$

$$< \infty,$$

since $\hat{\mu}_n(2\gamma) < \hat{\mu}_n(\gamma) < 1$ and $\text{E}[\hat{\xi}(\gamma)^2] < \infty$ and this ends the proof.   □

PROPOSITION (8.3)

Assume a well behaved branching process and let $\hat{\xi}^{(n)}(t)$ denote as usual the cardinality of the set of individuals in the nth generation by time t. Then with $\gamma < \alpha$ it can be shown that $\text{Var}[\hat{\xi}^{(n)}(\gamma)] < \infty$ if $\text{Var}[\hat{\xi}(\gamma)] < \infty$ and $\hat{\mu}(\gamma) < \infty$.

PROOF

To begin with we use the variance decomposition on $\hat{\xi}^{(n)}(\gamma)$ and get

$$\text{Var}[\hat{\xi}^{(n)}(\gamma)] = \text{E}[\text{Var}[\hat{\xi}^{(n)}(\gamma) | \mathscr{A}_{n-2}]] +$$
$$\text{Var}[\text{E}[\hat{\xi}^{(n)}(\gamma) | \mathscr{A}_{n-2}]],$$

where $\mathscr{A}_n$ is the σ–algebra generated by the lives of all individuals in $\overset{n}{\underset{k=0}{\cup}} N^k$, and $\mathscr{A}_{-1}$ is the trivial σ–algebra. Notice now that

$$\hat{\xi}^{(n)}(\gamma) = \underset{x \in N^n}{\Sigma} e^{-\gamma \sigma_x}$$
$$= \underset{x \in N^{n-1}}{\Sigma} \ \underset{k \in N}{\Sigma} e^{-\gamma(\sigma_x + \tau(k, \omega_x))}.$$

The first term of the variance decomposition becomes

$$E[\text{Var}[\sum_{x \in N^{n-1}} e^{-\gamma\sigma_x} \sum_{k \in N} e^{-\gamma(\sigma_x + \tau(k,\omega_x))} | \mathscr{A}_{n-2}]]$$

$$= E[\sum_{x \in N^{n-1}} e^{-2\gamma\sigma_x} \text{Var}[\sum_{k \in N} e^{-\gamma\tau(k,\omega_x)}]]$$

$$= \text{Var}[\hat{\xi}(\gamma)]E[\hat{\xi}^{(n-1)}(2\gamma)]$$

$$= \text{Var}[\hat{\xi}(\gamma)]\hat{\mu}(2\gamma)^{n-1},$$

where we use the branching property.

It is thus seen that since $\hat{\mu}(2\gamma) < \hat{\mu}(\gamma)$, the first term of the variance decomposition is finite if the requirements of the proposition are fullfilled. Returning to the second term, we use the same idea as above to get that it equals

$$\text{Var}[E[\sum_{x \in N^{n-1}} e^{-\gamma\sigma_x} \sum_{x \in N} e^{-\gamma\tau(k,\omega_x)} | \mathscr{A}_{n-2}]]$$

$$= \text{Var}[\sum_{x \in N^{n-1}} e^{-\gamma\sigma_x}]\hat{\mu}(\gamma)^2$$

$$= \text{Var}[\hat{\xi}^{(n-1)}(\gamma)]\hat{\mu}(\gamma)^2,$$

and the same procedure repeated $(n-1)$ times

$$= \text{Var}[\hat{\xi}(\gamma)]\hat{\mu}(\gamma)^{2(n-1)}. \qquad \qquad \square$$

It is now possible to formulate immediate corollaries giving $\tilde{E}[f(d)]$ and $\tilde{E}[f'(d)]$. In the case of $\tilde{E}[f']$ we get

COROLLARY (8.4)

$$\tilde{E}(f(d)] = \hat{\mu}_m(\alpha) \sum_{k=0}^{d} \int_0^\infty \alpha e^{-\alpha t} E[y^k(t)y^{d-k}(t)$$

$$- y^{k-1}(t)y^{d-k-1}(t)]dt < \infty$$

if the conditions of LEMMA (8.2) are satisfied.

Before we close this section, we turn our attention to another quantity that may be of interest in the context of our present model namely the number, $m(d)$, of the labels represented by at least one individual among those in the $F(d)$ family of the RSI. We are not going to work out the details since this situation is much alike the one considered in THEOREM (8.1), but we will briefly discuss what can be done and in what way.

It is not hard to realize that, what we in fact consider is the number of "relatives" of the label of the RSI, related to this by being its (b,f)–cousins, where b and f are such that $b+f = d$ and b,f in $\{0\} \cup N$. By (b,f)–cousins we simply mean those related to the label of the RSI by being the f–generation progeny of her bth ancestor. Another way of viewing this is to only consider the mutants among the individuals in the $F(d)$ family of the RSI.

With $\xi'^{(k)}$ meaning, as earlier, the kth generation labels, it can be shown that

$$\tilde{E}[m(d)] = \hat{\mu}_m(\alpha) \int_0^\infty \alpha e^{-\alpha t} \sum_{k=0}^{d} E[\xi'^{(k)}(t)\xi'^{(d-k)}(t) - \xi'^{(k-1)}(t)\, \xi'^{(d-k-1)}(t)]dt$$

and that this will be finite under the assumptions of THEOREM (8.1).

8.2. ANOTHER MEASURE OF RELATEDNESS

In Chapter Seven we introduced the probability of identity by descent as a measure of the overall variation over labels within the population. An obvious manner of generalizing this concept in the context of the present chapter, is to consider the probability of two randomly sampled individuals being at some given genetic distance. If this distance is taken as d, we talk about d−order identity by descent. Taking $d = 0$ will of course give the situation considered in THEOREM (7.4).

Let 0_1 and 0_2 denote a first and a second RSI chosen with replacement from an old well behaved labelled branching population. Let $f_{0_1}(d)$ be the cardinality of the F(d) family pertaining to 0_1. Let further P_t denote the conditional probability that 0_1 and 0_2 are d−order identical by descent given the branching population evolution. Then with Prob denoting that we calculate probabilities using the law induced by the sampling mechanism

$$P_t = Prob(0_2 \in F_{0_1}(d))$$

$$= \sum_{x \in I} Prob(0_2 \in F_{0_1} \mid 0_1 = x) \cdot Prob(0_1 = x)$$

$$\sim \frac{1}{y_t} \cdot \sum_{x \in I} \frac{f_x(d)}{y_t}.$$

CORROLLARY (8.5)

As $t \to \infty$, and under the conditions of THEOREM (8.1)

$$P_t \sim \tilde{E}[f(d)]/y_t.$$

REMARKS

—   [Chakraborty, 1975] has studied a similar measure of genetic variation in the context of another model.

—   As already mentioned it is possible to formulate counterparts of the results in this chapter in the case where sampling is restricted to the set of individuals still alive at the sampling time point (cf. Chapter Seven).

CHAPTER NINE

RELATED INDIVIDUALS

As pointed out in [Jacquard, 1974], the starting point of all genetical thought must have been the fact that related individuals resemble one another. A question that many geneticists ask is "what information about an individual x can we gain from information about another individual y related to x in a given manner"? From the point of view of the evolutionist it is relevant to ask the inverse question "How is x likely to be related to y, given some knowledge about the genetic constitution of x and y or at least about their relative genetic relationship"? It may, for example, happen that we are in possession of their complete nucleotide sequences and want to use this knowledge to make some inference about the backward time since the two lineages they belong to may have diverged.

9.1. GENEALOGICAL DISTANCES

Before we make any attempt to answer the above questions within the scope of our model, we have to make the concepts "related" and "resemble" more rigorous. Of course, the word related must be understood in terms of genealogical relationships such as parent–offspring, sibs, cousins, etc. [Jagers, 1981] is entirely devoted to the study of such relationships in the context of general branching processes. In this reference, an RSI is chosen and questions are asked about the number of individuals related to her in a specific manner. Here we adopt a slightly different point of view and assume that two RSI:s are chosen at random and ask questions about them being related in a specific manner. As was discussed in Chapter Five and Chapter Six, similar results as those obtained for two RSI:s can be proved when the sampling is restricted to those still alive at the sampling time point. More precisely we will consider the following measure, g, of what may be called genealogical distance between individuals x and y

$$g(x,y) = (b,f)$$

meaning that in order to go between x and y along their common family tree, the minimal path goes exactly b backward and f forward generations.

Obviously $\qquad\qquad\qquad\qquad g(y,x) = (f,b)$

and $\qquad\qquad\qquad\qquad g(x,y) = (0,1) <=> y_{[1]} = x.$

By individuals x and y being alike, we will simply mean that y belongs to some version of the F(k,i) family pertaining to x for some given values of k and i. Notice that this applies generally and not only in the infinite sites model. The reason is that we can always assume that every mutation leads one (or more) step(s) away from the ancestor, even though step cannot be identified with a difference with respect to one site. In other words we define a measure of the distance, $d(x,y)$, between any two individuals x and y by

$$d(x,y) = (k,j)$$

with the interpretation that along the minimal genealogical path connecting x and y exactly k backward and j forward mutations are to be encountered.

It may also be argued for the fact that in many cases it is relevant to only consider the first components of g and d. This will certainly be the case if we are interested in the time backwards since divergence of the lines that the two randomly sampled individuals belong to. Obviously the probabilities of identity by descent between two RSI:s and more generally the probability that they belong to the same $F(j)$ family, for some j given that they are related in a specific manner can be obtained as special cases.

9.2.   DISTANCES BETWEEN RELATIVES

Assume that two RSI:s 0_1 and 0_2 are chosen among all those born in a well behaved branching population at some late time moment t. As was discussed in the introduction to the present chapter, it might be of interest to consider the probabilities

(1) $P_t(d(0_1, 0_2) = (j,k)|g(0_1, 0_2) = (b,f))$

(2) $P_t(g(0_1, 0_2) = (b,f)|d(0_1, 0_2) = (j,k))$.

But

$$(1) \quad = \frac{P_t(d(0_1,0_2)=(j,k); \; g(0_1, 0_2) = (b,f))}{P_t(g(0_1, 0_2) = (b,f))}$$

$$= \frac{\displaystyle\sum_{\substack{x \in I \\ \sigma_x \le t}} P_t(d(0_1,0_2) = (j,k); g(0_1,0_2) = (b,f); 0_1 = x)}{\displaystyle\sum_{\substack{x \in I \\ \sigma_x \le t}} P_t(g(0_1,0_2) = (b,f); 0_1 = x)}$$

$$= \frac{\displaystyle\sum_{\substack{x \in I \\ \sigma_x \le t}} \#\{y \in I ; \; \sigma_y \le t; \; d(x,y) = (j,k); \; g(x,y) = (b,f)\}/y_t}{\displaystyle\sum_{\substack{x \in I \\ \sigma_x \le t}} \#\{y \in I; \; \sigma_y \le t; \; g(x,y) = (b,f)\}/y_t}.$$

Fortunately this can be written in the form

$$z_t^{\chi_1}/z_t^{\chi_2}$$

for some random characteristics χ_1 and χ_2. The idea now is that once this is done, the basic convergence theorems will show that $(1) \to E[\hat{\chi}_1(\alpha)]/E[\hat{\chi}_2(\alpha)]$, in some suitable sense.

It is however to be noted that both $Z_t^{X_1}/y_t$ and $Z_t^{X_2}/y_t$ are empirical ratios of the type whose asymptotics can be studied using the stable pedigree measure $\tilde{P}$. In order to do so, one has to think of them as $\tilde{P}$–expectations of functionals of the pedigree of the RSI. More precisely $\lim_{t \to \infty} Z_t^{X_2}/y_t$ corresponds to $\tilde{E}[\varphi_2]$ where φ_2 counts the cardinality of the set of her (b,f) cousins belonging to her (j,k) family.

Formally

$$\tilde{E}[\varphi_2] = \tilde{E}[\xi_{-b}^{(f)}(-\tilde{\sigma}_{-b}) - \xi_{-b+1}^{(f-1)}(-\tilde{\sigma}_{-b+1})]$$

which, using LEMMA (5.1) yields

$$\tilde{E}[\varphi_2] = \sum_{x \in N^b} E[e^{-\alpha\sigma_x} \int_0^\infty \alpha e^{-\alpha t} \xi^{(f)}(t+\sigma_x) dt]$$

$$- \sum_{x \in N^{b-1}} E[e^{-\alpha\sigma_x} \int_0^\infty \alpha e^{-\alpha t} \xi^{(f-1)}(t-\sigma_x) dt]$$

$$= \int_0^\infty \alpha e^{-\alpha t} E[\xi^{(b)}(t)\xi^{(f)}(t) - \xi^{(b-1)}(t)\xi^{(f-1)}(t)] dt.$$

If we now let $\zeta_j^{(f)}(\cdot)$ stand for the cardinality of the fth generation progeny of the ancestor with exactly j intervening mutant ancestors (recall that an individual can be interpreted as her own ancestor!), then proceeding as above

$$\tilde{E}[\varphi_1]$$

$$= \hat{\mu}_m(\alpha) \int_0^\infty \alpha e^{-\alpha t} E[\zeta_j^{(f)}(t)\zeta_k^{(b)}(t) - \zeta_{j-1}^{(f-1)}\zeta_{k-1}^{(b-1)}(t)] dt.$$

Naturally this can also be derived using a slightly modified version of the random characteristic used in the proof of THEOREM (8.1). We are now ready to state

THEOREM (9.1)

Consider a well–behaved branching process satisfying (C.7), and let everything be as above. Then

(a) $P_t(d(0_1,0_2): (j,k)|g(0_1,0_2) = (b,f))$

$$\longrightarrow \frac{\hat{\mu}_m(\alpha) \int_0^\infty \alpha e^{-\alpha\mu} E[\zeta_j^{(f)}(u)\zeta_k^{(b)}(u) - \zeta_{j-1}^{(f-1)}(u)\zeta_{k-1}^{(b-1)}(u)] di]}{\int_0^\infty \alpha e^{-\alpha u} E[\xi^{(f)}(u)\xi^{(b)}(u) - \xi^{(f-1)}(u)\xi^{(b-1)}(u)] du}$$

as $t \to \infty$ in probability on $\{y_t \to \infty\}$.

(b) $\qquad P_t(g(0_1,0_2) = (b,f)|d(0_1,0_2) = (j,k))$

$$\longrightarrow \frac{\int_0^\infty \alpha e^{-\alpha t} E[\zeta_j^{(f)}(u)\zeta_k^b(u) - \zeta_{j-1}^{(f-1)}(u)\zeta_{k-1}^{(b-1)}(u)]\,du}{\hat{\mu}_m(\alpha)\int_0^\infty \alpha e^{-\alpha u} E[y^i(u)y^k(u) - y^{i-1}(u)y^{k-1}(u)]\,du}$$

as $t \to \infty$ in probability on $\{y_t \to \infty\}$.

The proof and the exact conditions for the convergence and the finiteness of the limits in (a) and (b) being much like those of the preceding chapter, will be omitted.

REMARKS

— It is possible to derive a recursion formula for the computation of $E[\xi^{(f)}(t)\xi^{(b)}(t)]$ according to

$$E[\xi^{(f)}(t)\xi^{(b)}(t)] = E[\sum_{i,j} \xi^{(f-1)}(t-\tau(i))\xi^{(b-1)}(t-\tau(j))]$$

$$= \int_{0\le u\ne v} \mu^{*(f-1)}(t-u)\mu^{*(b-1)}(t-v)E[\xi(du)\xi(dv)]$$

$$+ \int_0^t E[\xi^{(f-1)}(t-u)\xi^{(b-1)}(t-u)]\mu(du).$$

— As in Chapter Seven it is straightforward to replace the two RSI:s by two randomly sampled individuals chosen among those alive at the sampling time point.

CHAPTER TEN

EMBEDDINGS IN BRANCHING PROCESSES

Some aspects of branching processes are better understood in terms of suitable groups of individuals rather than in terms of the individuals themselves. A convenient name which may be given to such groups is generalized individuals. The label process studied in Chapter Four is but one example of a situation where it is fruitful to think in terms of such generalized individuals. Other such examples can be found in [Doney, 1976] and in [Broberg, 1987].

Sometimes a population of generalized individuals may be viewed as a branching population which is embedded in the underlying one. When this is the case we can apply standard branching process theory on the corresponding embedded branching process. As a biproduct it turns out that it is possible to consider certain processes counted by non–individual random characteristics as being counted by individual characteristics on the new level (cf. the discussion of $\text{Var}[M_t]$ at the end of §4.1).

We begin by making available some pertinent concepts and terminology as a background against which to view generalized individuals. More specifically we will present branching processes defined on spaces of marked trees as they are introduced in [Neveu, 1986] and further explored in [Chauvin, 1986]. After having explained the relation between this new construction of a branching process and the old one, we will describe a device yielding groups of individuals. We will also show that a population consisting of such groups will possess the branching property as it is formulated on a space of marked trees.

The main reason of going via spaces of marked trees is that our old construction of a branching process is "too large" (contains coordinates belonging to never born individuals), and it is not very clear how to enlarge the embedded marked trees of generalized individuals in order to recover the branching process characterization we have used until now.

10.1. TREES AND GENERAL BRANCHING PROCESSES

let I stand as usual for the individual space. A subset u of I will be called a tree if $0 \in u$ and $x \in u \Rightarrow x_{[1]} \in u$. The class U of all trees in I will be referred to as the tree space, and the symbol U_x will be reserved to denote the subset of U defined by $U_x = \{u \in U; x \in u\}$. Then certainly $U_0 = U$. For every individual $x \in u$, we let $T_x(u)$ denote the translated tree, obtained by translating the tree u to its xth component. T_x is thus a mapping $T_x: U_x \to U$. The word translation is used here to emphasise the fact that T is a translation operator. The tree space U is equipped with the σ-algebra $\mathscr{F} = \sigma(U_x; x \in I)$. Recall now that Ω stands for the life space on which we make the usual assumptions (i.e. we assume it to be large enough to contain information about life span, reproduction etc). Naturally Ω can also be viewed as a sort of mark space in the terminology of [Neveu, 1986].

Our new sample space will now be a space, U^Ω, of marked trees, where a marked tree, $\bar{u} \in U^\Omega$, is specified by giving a tree $u \in U$ and a life history (mark), ω_x, for every $x \in u$. Formally we can write $\bar{u} = (u, (\omega_x; x \in u))$. If we let p denote the canonical projection mapping $p: U^\Omega \to U$ associating with every marked tree, $\bar{u}$, the tree, u, obtained by disregarding the marks, then $\omega_x; x \in I$ define coordinate projection mappings

$$\omega_x: U^\Omega_x = p^{-1}(U_x) = \{\bar{u}; x \in u\} \to \Omega.$$

The space, U^Ω, of marked trees can be equipped with the σ-algebra

$$\mathscr{F}^\Omega = \sigma(p^{-1}(\mathscr{F}), \omega_x; x \in I).$$

Also translation mappings $T^\Omega_x: U^\Omega_x \to U^\Omega$ can be defined through

$$T^\Omega_x(\bar{u}) = (T_x(u), (\omega_{xy}; y \in T_x(u))) \quad \text{when } \bar{u} = (u, (\omega_x; x \in u)).$$

Of course, these translation mappings are related to T_x through

$$T_x \circ p = p \circ T^\Omega_x$$

on U^Ω_x. Now $\omega_x, x \in I$ can be seen as $\omega_0 = \omega$ translated by T^Ω_x, i.e.

$$\omega_x = \omega \circ T^\Omega_x.$$

As before different aspects of the life of an individual, $x \in I$, can be defined as measurable mappings but now on U^Ω_x. Just in order to illustrate this consider

$$\lambda_x: U^\Omega_x \xrightarrow[T^\Omega_x]{} U^\Omega \xrightarrow{\omega} \Omega \xrightarrow{\lambda} R^+,$$

i.e.

$$\lambda_x = \lambda \circ \omega \circ T^\Omega_x = \lambda \circ \omega_x$$

on U^Ω_x.

Let $\omega^* \in \Omega^I$ denote an outcome of a general branching process and consider the mapping $\psi: \Omega^I \to U^\Omega$ defined formally by

$$\psi(\omega^*) = (\{x: \sigma_x < \infty\}, (\omega_x; x \in \{y; \sigma_y < \infty\})), \; \omega^* \in \Omega^I.$$

In words, the mapping ψ associates the marked tree $\psi(\omega^*)$ with every $\omega^* \in \Omega^I$, in such a manner that $\psi(\omega^*)$ contains precisely the individuals x for which $\sigma_x(\omega^*) < \infty$ (the set of such individuals being obviously a tree) and their usual life coordinates as marks. Now ψ induces a probability measure, $P^I\psi^{-1}$, on U^Ω for any life law P on Ω. A general branching process with life law P is then, in the sense of [Neveu, 1986], the triplet $(U^\Omega, \mathscr{A}^\Omega, P^I\psi^{-1})$. Also the translation mappings $T_x^\Omega: U_x^\Omega \rightarrow U^\Omega$ and the usual shifts $S_x: \Omega^I \rightarrow \Omega^I$ are related by

$$T_x^\Omega \circ \psi = \psi \circ S_x$$

on $\psi^{-1}(U_x^\Omega)$. The precise relation between the two constructions is contained in the following proposition which is a very slight modification of Proposition (3) in [Neveu, 1986].

PROPOSITION (10.1)

With everything defined as above, and for every life law P on Ω, $Q = P^I\psi^{-1}$ is the unique probability law on $(U^\Omega, \mathscr{A}^\Omega)$ which is such that the life $\omega = \omega_0$ of the ancestor follows P and T_i^Ω, $i = 1,..., j$, when $\xi(\omega) = j$ are independent and identically distributed according to Q conditionally on ω.

PROPOSITION (10.1) can be said to contain the essence of the branching property (see below). The proof provided in [Neveu, 1986] works here (as is also noticed at the end of that reference) mutatis mutandis. The rather important uniqueness (not discussed in that proof) is a consequence of the fact that this branching property characterization actually fixes any "cylinder probability"

$$Q(\{\bar{u} \in U^\Omega; x_1 \in p(\bar{u}), ... , x_1 x_2 \cdots x_k \in p(\bar{u}),$$

$$x_1 \cdots x_{k+1} \notin p(\bar{u}),...,x_1 \cdots x_{k+n} \notin p(\bar{u}),$$

$$\omega_{x_1} \in A_1 ,..., \omega_{x_1 \cdots x_k} \in A_k\})$$

for any $k, n, x_1,..., x_{k+n}, A_1,..., A_k$, which can be seen by an iterative alternating use of the required first generation branching property and the translation operators. It is thus seen that as long as we do not involve the non−realized individuals (i.e. individuals $x \in I$ such that $\sigma_x(\omega^*) = \infty$) the two constructions are equivalent. In other words, studying Ω^I equipped with S_x and P^I is essentially the same thing as studying U^Ω equipped with T_x^Ω and the Q corresponding to P.

10.2. THE BRANCHING PROPERTY

Having discussed the relation between the two sample spaces of a branching process, our aim will now be to describe a generalisation of the branching property on trees. This will be later on combined with the characterization of Q in PROPOSITION

(10.1) to study the embedded processes of generalized individuals. The above mentioned enlargement of the branching property was first formulated in [Chauvin, 1986]. We begin by introducing some notation.

Let $>$ denote the obvious partial order relation on I defined by $x > y <=> \exists$ $k \in N; x_{[k]} = y$, simply saying that x stems from y. Of course then $y < x$ means that y is an ancestor of x. If M and L are subsets of I, then $L > M$, means that $\forall x \in L$, $\exists y \in M$ such that $x > y$. If for some x and y in I neither $x > y$ nor $y > x$ is satisfied we agree to say that x and y are noncomparable. Of special interest will be the subsets, L, of I meeting the following requirement

$$x \in L \ \& \ y \in L => x \text{ and } y \text{ are noncomparable.}$$

We follow [Chauvin, 1986] in calling such subsets stopping lines. In the same reference are also introduced stopping operators M_L corresponding to stopping lines L, defined on U^Ω by taking $M_L(\bar{u})$ to be the "stopped" marked tree obtained from $\bar{u}$ by removing from this the coordinates of the individuals stemming from L. Using M_L we can define the pre–L σ–algebra $\mathscr{A}_L^\Omega$ by $\mathscr{A}_L^\Omega = M_L^{-1}(\mathscr{A}^\Omega)$ (for details cf. the above reference). Since

$$L' < L => \mathscr{A}_{L'}^\Omega \subset \mathscr{A}_L^\Omega,$$

$(\mathscr{A}_L^\Omega)_{L \in 2^I}$ is a filtration.

Take J to be a measurable mapping $U^\Omega \to 2^I$ which is such that $J(\bar{u}) \subset p(\bar{u})$, and $\{\bar{u} \in U^\Omega; J(\bar{u}) < L\} \in \mathscr{A}_L^\Omega$. If J is also a stopping line with probability one, it will be called an optional line. This means that J is determined by its ancestors and their side branches (for details cf. [Jagers, 1989]). If J is an optional line we define

$$\mathscr{A}_J^\Omega = \{A \in \mathscr{A}^\Omega; A \cap \{L > J\} \in \mathscr{A}_L \ \forall L \subset I\}.$$

We are now able to state the following result which is exactly the tool we need in the next subsection. In [Chauvin, 1986] this result is given for fixed finite stopping lines on a space of trees. In [Jagers, 1989] it is given in terms of an arbitrary optional line on $(\Omega^I, \mathscr{A}^I, P^I)$. Reconciling the two formulations so as to obtain it for arbitrary optional lines on trees results in

PROPOSITION (10.2)

Let J be an optional line. Then conditionally with respect to $\mathscr{A}_J^\Omega$ the translated trees T_x^Ω, $x \in J$ are independent and identically distributed according to Q. This means that for all real, positive and measurable functions f_x, $x \in I$

$$E_Q[\prod_{x \in J} f_x \circ T_x^\Omega | \mathscr{A}_J^\Omega] = \prod_{x \in J} E_Q[f_x].$$

where E_Q denote expectation using Q as probability measure.

PROOF

The statement of proposition being proved on Ω^I by [Jagers, 1987], the technique of [Neveu, 1986] also used in [Chauvin, 1986] will cooperate nicely to prove the proposition.   □

10.3. GENERALIZED INDIVIDUALS

We are now in position to discuss how the above introduced tools can be put together to construct embedded processes of generalized individuals. This will be done on U^Ω which, we remind, was obtained from our basic outcome space Ω^I through the mapping ψ.

For this purpose we fix an optional line $\phi : U^\Omega \to 2^I$ as was described in the preceding subsection. ϕ is now intended to be used according to the following scheme. To the group of individuals stemming from the ancestor but not from ϕ, we will associate a hypothetical ancestor, $\mathbf{0}$. The life (mark) of this generalized ancestor will be the "stopped" marked tree consisting of the individuals in the group together with their usual life coordinates. To every $x \in \phi$, we think of a realized generalized child of $\mathbf{0}$, whose life is again a stopped marked tree, namely the one consisting of the individuals stemming from x but not from $\phi_x = \phi \circ T_x^\Omega$ together with their life coordinates. It turns out that repeating this procedure yields a new space of marked trees which we formally describe and show to possess the required branching character in what remains of this subsection.

Let U^E stand for the new (embedded) space of marked trees (see below) and let $\tilde{u}$ be an element of U^E. It is convenient to think of a mapping $\Lambda : U^\Omega \to U^E$ giving a marked tree $\tilde{u} = \Lambda(\bar{u})$ for every outcome $\bar{u} \in U^\Omega$. In what follows we will give a full description of U^E and discuss how Λ operates on U^Ω .

Since a marked tree $\tilde{u} = \Lambda(\bar{u}) \in U^E$ is specified by giving a tree $p(\tilde{u})$ and marks ω'_x, $x \in p(\tilde{u})$ (in the sequel we will use bold characters to denote generalized individuals and ω' to denote their life coordinates), the mapping Λ can be describe sequentially.

–   The ancestor 0 of the underlying population will be called a line–initiator, meaning that she initiates a group of individuals to be thought of as the generalized individual **0**, the new ancestor. In the sequel the term line–initiator will be reserved to denote individuals initiating groups of individuals to be viewed as generalized individuals.

Now since $0 \in p(\bar{u})$ $\forall \bar{u}$, this will imply that $0 \in p(\tilde{u})$, $\tilde{u} = \Lambda(\bar{u})$ for every outcome $\bar{u}$.

The life (mark) of the new ancestor, denoted by $\omega_0' = \omega'$, will be taken to be the stopped marked tree $M_\phi(\bar{u})$.

—   The first generation indivduals in $p(\tilde{u})$, i.e. the offspring of $\mathbf{0}$, are specified by means of what follows: The individuals of ϕ are line–initiators (cf. above). They can always be ordered according to their birth moments (and according to some arbitrary but fixed rule if those coincide). One can now think of the individual ranked as number i among the other indivduals in ϕ as the line–initiator of the group of individuals to be identified with, $\mathbf{i}$, the ith child of $\mathbf{0}$. From a formal point of view this will result in $\mathscr{I}_\phi^\Omega$ — measurable mappings which we, for notational convenience (cf. below), write $i(\omega')$ to be thought of as the individual ranked as number i among the individuals in ϕ as discussed above. $i(\omega')$ is thus a mapping taking values in I and $i = 1,2,..., \nu(\omega')$ where $\nu(\omega')$ is the mapping giving the cardinality of ϕ. The rationale of the above notation is that since ϕ depends on the outcome, $\bar{u}$, of the process only through $M_\phi(\bar{u}) = \omega'(\bar{u}) = \omega'$, it is itself determined by this stopped tree once the outcome $\bar{u}$ is fixed. Using this notation, it is seen that

$$i \in p(\tilde{u}) <=> i \le \nu(\omega')$$

and

$$\tau'(i, \omega') = \sigma_{i(\omega')}$$
$$= \text{the birth time of the ith generalized child of } \mathbf{0},$$

$i = 1,... \nu(\omega')$.

As usual, this defines a reproduction point process which we denote ξ'. Naturally then $\xi'(\infty) = \nu(\omega')$.

—   The lives of the first generation generalized individuals will be defined by $\omega_i' = \omega'$ $\circ T_{i(\omega')}^\Omega(\bar{u})$ when $i \le \nu(\omega')$.

—   Proceeding as above yields

$$\mathbf{xi} \in p(\tilde{u}) <=> \mathbf{x} \in p(\tilde{u}) \,\&\, i \le \nu(\omega_\mathbf{x}')$$

where $\omega_\mathbf{x}'$ is $\mathbf{x}$'s life and, for $\mathbf{xi} \in p(\tilde{u})$,

$$\omega_{\mathbf{xi}}' = \omega' \circ T_{\mathbf{x}_1(\omega')\mathbf{x}_2(\omega_{\mathbf{x}_1}') \cdots \mathbf{x}_{n(\mathbf{x})}(\omega_{\mathbf{x}_{[1]}}')i(\omega_\mathbf{x}')}^\Omega(\bar{u}).$$

Clearly this describes both the new space U^E and the mapping Λ by means of which it is obtained. Moreover we have a hint about the nature of the translation mappings in the new space. These will be mappings $T_\mathbf{x}^E : U_\mathbf{x}^E \to U^E$ related to the translation mappings in U^Ω through the following: Let x stand for the line initiator "generating" the generalized individual $\mathbf{x}$, i.e.

$$x = \mathbf{x}_1(\omega')\mathbf{x}_2(\omega'_{\mathbf{x}_1}) \cdots \mathbf{x}_{n(\mathbf{x})}(\omega'_{\mathbf{x}_{[1]}})$$

(this comfortable convention will be used in the sequel without further explanation). Then

$$T^E_{\mathbf{x}}(\Lambda(\overline{u})) = \Lambda(T^\Omega_{\mathbf{x}}(\overline{u})).$$

We thus see that for every outcome, ω^*, of the basic outcome space Ω^I, a generalized individual is an element of $p(\tilde{u})$ where we use that $\tilde{u} = \Lambda(\overline{u})$ and $\overline{u} = \psi(\omega^*)$, and that the life of a generalized individual, $\mathbf{x}$, is only defined on $U^E_{\mathbf{x}} = \{\tilde{u}; \mathbf{x} \in p(\tilde{u})\}$.

Let us now turn our attention to the probability space $(U^E, \mathscr{F}^E, P)$, which is induced from $(U^\Omega, \mathscr{F}^\Omega, Q)$ by Λ. The new probability measure P is defined by

$$P(A) = Q\Lambda^{-1}(A)$$
$$= Q(\{\overline{u}; \Lambda(\overline{u}) \in A\}),$$

$A \in U^E$. Notice that the mapping Λ is such that it preserves all the information available in the old process and that it is easily seen to be one to one. It thus follows that if f is any measurable real valued function on U^E, there will exist a real valued function f' on U^Ω defined by $f'(\overline{u}) = f(\Lambda(\overline{u}))$ or equivalently $f(\tilde{u}) = f'(\Lambda^{-1}(\tilde{u}))$. This fact will be crucial in proving the promised branching property of the embedded process of generalized individuals corresponding to the fixed optional line ϕ.

THEOREM (10.3)

Let $\mathbf{N}$ denote the (random) set of first generation generalized individuals, and let $f_{\mathbf{x}}$, $\mathbf{x} \in I$ be integrable real valued functions on $U^E_{\mathbf{x}}$. Then the following holds with probability one

$$E[\prod_{\mathbf{x} \in \mathbf{N}} f_{\mathbf{x}} \circ T^E_{\mathbf{x}} | \mathscr{F}^E_{\mathbf{N}}]$$
$$= \prod_{\mathbf{x} \in \mathbf{N}} E[f_{\mathbf{x}}]$$

where E denotes expectation w.r.t. P.

PROOF

Use the fact that

$$f_{\mathbf{x}} \circ T^E_{\mathbf{x}}(U^E_{\mathbf{x}}) = f'_{\mathbf{x}} \circ T^\Omega_{\mathbf{x}} \circ \Lambda^{-1}(U^E_{\mathbf{x}})$$
$$= f'_{\mathbf{x}} \circ T^E_{\mathbf{x}}(U^\Omega_{\mathbf{x}})$$

as is seen from the discussion preceding the theorem. Now since $f_{\mathbf{x}} \circ T^E_{\mathbf{x}}$ acting on $U^E_{\mathbf{x}}$ does the same thing as $f'_{\mathbf{x}} \circ T^\Omega_{\mathbf{x}}$ acting on $U^\Omega_{\mathbf{x}}$, by the very definitions of P and $\mathscr{F}^E_{\mathbf{x}}$, the theorem will follow if we can prove that (with probability one)

$$E_Q[\prod_{x;\,x\in N} f'_x \circ T^\Omega_x \mid \mathscr{F}^\Omega_{\{x;\,x\in N\}}]$$

$$= \prod_{\{x\,;\,x\in N\}} E_Q[f'_x].$$

Naturally $\{x;\, x \in N\} = \phi$ and the above equality reduces to

$$E_Q[\prod_{x\in\phi} f'_x \circ T^\Omega_x \mid \mathscr{F}^\Omega_\phi]$$

$$= \prod_{x\in\phi} E[f'_x],$$

which is an immediate consequence of PROPOSITION (10.2) since ϕ is an optional line. □

As already mentioned, the uniqueness of the measure Q on $(U^\Omega, \mathscr{F}^\Omega)$, implies that PROPOSITION (10.1) contains the essence of the branching character. Since the above theorem shows that the (generalized) translated trees possess the conditional independence property described in that proposition, we can conclude that the embedded process of generalized individuals will behave as a branching process. A less direct approach leading to the same conclusion is to consider an arbitrary optional stopping line of generalized individuals and to show the counterpart of PROPOSITION (10.2) now on the level of generalized individuals. The proof of such a result will be almost identical with the proof of THEOREM (10.3). Once proved this result will imply the validity of the branching lemmas in [Jagers & Nerman, 1984a]. Since those lemmas are the essential ingredients of the proofs of the basic convergence theorems, those theorems can be used on the level of the embedded processes.

Naturally, counterparts of different kinds of entities and quantities associated with branching processes can be defined for the embedded processes.

$$\sigma_{\mathbf{x}} = \sigma_x$$

$$\lambda_{\mathbf{x}} = \max\{(\lambda_{xy} + \sigma_{xy});\, y \in \mathbf{x}\}$$

etc.

Also random characteristics can be defined as $\mathscr{B} \times \mathscr{F}^E$ — measurable mappings $\chi: R^+ \times U^E \to R^+$. For an arbitrary generalized individual $\mathbf{x}$, $\chi_{\mathbf{x}}$ is mapping $R^+ \times U^E_{\mathbf{x}} \to R^+$. This can be used to define branching processes, $\mathbf{Z}^\chi_t$, counted by random characteristics.

$$\mathbf{Z}^\chi_t(\tilde{u}) = \sum_{\mathbf{x}\in p(\tilde{u})} \chi_{\mathbf{x}}(t - \sigma_{\mathbf{x}}).$$

REMARKS

— It is straightforward to formulate a multitype version of the results above and thus cover the situation considered in [Doney, 1976].

— If χ is a usual random characteristic, then the process counted by this characteristic can be written as $Z_t^\chi(\overline{u}) = \sum\limits_{x \in p(\overline{u})} \chi_x(t-\sigma_x)$ or in terms of generalized individuals as (with $\tilde{u} = \Lambda(\overline{u})$ and $L_x(\overline{u}) = p(M_\phi \circ T_x^\Omega(\overline{u}))$).

$$\mathbb{Z}_t^\chi = \sum_{x \in p(\tilde{u})} \sum_{y \in L_x} \chi_y(t-\sigma_y)$$

$$= \sum_{x \in p(\tilde{u})} \chi_x(t-\sigma_x)$$

$$= \mathbb{Z}_t^\chi$$

where

$$\chi_x(t-\sigma_x) = \sum_{y \in L_x} \chi_y(t-\sigma_y).$$

This implies that whenever it is convenient to regroup the individuals into generalized individuals, one can study $Z_t^\chi(\overline{u})$ through $\mathbb{Z}_t^\chi(\tilde{u})$. Examples of this are given in Chapter Four (cf. also the proof of the main theorem in Chapter Eight).

CHAPTER ELEVEN

EXTENSIONS

In this chapter we present some extensions of the results presented in the preceding chapters as well as some suggestions for further research.

11.1. OTHER SAMPLING SCHEMES

A first extension of the sampling schemes used so far is to assume that we sample among individuals in some other subcategory of those born at or before the sampling time point than those living.

But there are other sampling aspects that are natural to think of. If we for instance sample not among individuals, but among labels, and let every label that has been seen in the population be chosen with a probability which is equal to the size of this label divided by the total number of individuals, and associate with each label a contribution which is again equal to its size, then clearly, the empirical mean of such sizes will be given by

$$\sum_{x \in I} 1(x \text{ mutant}) \ y_n(t - \sigma_x) \circ S_x \ \frac{y_n(t - \sigma_x) \circ S_x}{y_t}$$

which of course will again converge to

$$\hat{\mu}_m(\alpha) \int_0^\infty \alpha e^{-\alpha t} E[y_n^2(t)] dt$$

in some suitable sense, according to some suitable basic convergence theorem, under some suitable conditions. This is an example of what may be called biased sampling or weighted sampling.

One could further assume that the sampling of 0_1 and 0_2 in §7.3 is not performed at the same time point, but that we first choose 0_1 at t and then choose 0_2 at $t+u$. In this case, the probability of identity by descent is given according to

$$\text{Prob}(0_2 \in F_{0_1}) = \sum_{x \in I} \text{Prob}(0_2 \in F_{0_1} \mid 0_1 = x) \text{Prob}(0_1 = x)$$

$$= \frac{1}{Z_t} \sum_{x \in I} \text{Prob}(0_2 \in F_x)$$

$$= \frac{1}{Z_t} \sum_{x \in I} \frac{\varphi(\Pi_x, t + u - \sigma_x)}{Z_{t+u}}$$

$$= \frac{1}{Z_{t+u}} \sum_{x \in I} \frac{\varphi(\Pi_x, t - \sigma_x + u)}{Z_t}$$

$$\simeq \frac{1}{Z_{t+u}} \tilde{E}_\ell[\varphi(\tilde{\omega}, a+u)]$$

where $\tilde{E}_\ell[\varphi(\tilde{\omega}, a+u)]$ is as in Chapter Seven

11.2. FURTHER DEVELOPMENTS

The results in this monograph can be extended in various directions. Some of those are discussed in what follows:

A first development would be the formulation of critical, subcritical as well as "multitype" counterparts of the ideas dealt with in the single–type supercritical case. Naturally, new typical problems will arise. At least in the critical Markovian case, it will perhaps be possible to obtain similar results as those obtained in the scope of other models of population development as for instance the Wright–Fisher model or Moran's model (cf. [Ewens, 1979]).

Another extension that should be discussed is the analogous of the celebrated Ewens sampling formula (cf. [Ewens, 1990]). The idea here is to sample a finite number, n, of individuals and say something about the multidimensional distribution of the label configuration $(a_1, a_2,..., a_n)$ where a_i stands for the number of labels represented by exactly i individuals among those in the sample. It turns out that this cannot be carried out in a straightforward manner using random characteristics. Some insight can however be gained from the following special case: Assume that we sample three individuals $0_1, 0_2$ and 0_3 among all those born at or before some late time point t (both dead and living), the sampling being performed with replacement. Three possibilites arise

(i) $\qquad a_3 = 1$ and $a_i = 0 \; \forall i \neq 3$.

(ii) $\qquad a_1 = 1$ and $a_2 = 1$ while $a_i = 0 \; \forall i \neq 1,2$

(iii) $\qquad a_1 = 3$ and $a_i = 0 \; \forall \; i \neq 1$.

Although the situation in (i) can be studied by means of random characteristics, we will prefer the following heuristic argument. Think of an urn–ball model where we have $N_t = \hat{\mu}_m(\alpha)y_t$ urns (one urn for each label) and recall from Chapter Seven that the asymptotic distribution of the number of individuals (balls) carrying the same label (from the same urn) as the RSI is

$$P_j = \hat{\mu}_m(\alpha)j \int_0^\infty \alpha e^{-\alpha u}P(y_n(u) = j)du, \quad j = 1, 2,... \; .$$

Now the probability of choosing a ball from the same urn three consecutive times (with replacement) is

$$P_t(a_3 = 1)$$

$$= \sum_j P_t(a_3 = 1, \text{ the urn in question contains } j \text{ balls})$$

$$= \sum_j P_t(a_3 = 1)|\text{the urn in question contains } j \text{ balls})$$

$$\qquad P_t(\text{the urn is question contains } j \text{ balls})$$

$$- \sum_j \frac{j^2}{y_t^2} \hat{\mu}_m(\alpha) j \int_0^\infty \alpha e^{-\alpha u} P(y_n(u) = j) du$$

$$= \frac{\hat{\mu}_m(\alpha)}{y_t^2} \int_0^\infty \alpha e^{-\alpha u} E[y_n(u)^3] du.$$

As already mentioned the above approximation can be derived properly using random characteristics. Before discussing how this can be used to say something about the general case let us notice that clearly the expression

$$\int_0^\infty \alpha e^{-\alpha u} E[y_n(u)^3] du$$

will be finite if there exists some $\gamma < \frac{\alpha}{3}$ meeting the following conditions

1) $E[\hat{\xi}_n(\gamma)] < 1.$

2) $E[\hat{\xi}_n(\gamma)^3] < \infty.$

(conditions for the finiteness of higher integral moments of branching processes counted by random characteristics can be obtained using cumulant methods as is shown in [Jönsson, 1990]). Notice now that the above heuristic urn–ball argument can be extended to the case where we sample n individuals, yielding

$$P_t(a_n = 1, a_i = 0 \ \forall i \neq n)$$

$$- \frac{\hat{\mu}_m(\alpha)}{y_t^{n-1}} \int_0^\infty \alpha e^{-\alpha u} E[y_n(u)^n] du.$$

Even in this case we can use random characteristic–arguments. Unfortunately both the convergence and the direct Riemann integrability will necessitate the existence of higher moments of the y_n process. The idea is now that once we have done that, we can consider other label configurations using urn–ball arguments and knowing that nothing "worse" than the situation discussed above can happen. Returning to the case where n=3 we see that

$$P_t(a_2 = 1, a_1 = 1)$$

$$P_t(a_2 = 1, a_1 = 1)$$

$$= 3 P_t(\text{the first two chosen balls are from one urn}$$

$$\text{while the third is from another})$$

$$- 3 \sum_j \frac{j}{y_t} (1 - \frac{j}{y_t}) \hat{\mu}_m(\alpha) j \int_0^\infty \alpha e^{-\alpha u} P(y_n(u) = j) du$$

$$= \frac{3\hat{\mu}_m(\alpha)}{y_t} \left[\int_0^\infty \alpha e^{-\alpha u} \sum_j j^2 P(y_n(u) = j) du \right.$$

$$-\int_0^\infty \alpha e^{-\alpha u} \sum_j j^3 P(y_n(u) = j)du/y_t\Bigg]$$

$$= \frac{3\hat{\mu}_m(\alpha)}{y_t}\int_0^\infty \alpha e^{-\alpha u}E[y_n(u)^2]du - \frac{3\hat{\mu}_m(\alpha)}{y_t^2}\int_0^\infty \alpha e^{-\alpha u}E[y_n(u)^3]du.$$

Obviously some of the steps of the above reasoning are quite unorthodox but the purpose was to gain insight in how to think about the problem.

Another topic that is of interest is modelling electrophoretic experiments: Up to now, it has been implicitly assumed that it is perfectly possible to discriminate between distinct labels. In genetical applications however, distinction between labels is often made by means of gel electrophoresis (cf. [Wallace, 1981] Ch.4). More precisely assume that the best we can do is to assign to each individual a charge in the form of an integer valued random variable (with the risk that two individuals carrying two distinct labels may be assigned the same charge), and that every mutation leads to one of

(i)   A unit increase of the charge can be detected.

(ii)   A unit decrease of the charge can be detected.

(iii)   No effect on the charge level.

If we now further assume that the events described in (i)–(iii) occur or not depending on the life of the mother of the individual in question in some convenient manner, this will correspond to

$$\xi_m(t) = \xi_{m^+}(t) + \xi_{m^-}(t) + \xi_{m^0}(t).$$

Such mutations are usually called stepwise mutations. Naturally the assumption that every mutant initiates an entirely new label is now removed and recurrent mutations are allowed. A first model adopting this point of view goes back to [Kimura & Ohta, 1973]. Since then many attempts to construct such models have been done. In the literature they are known as, the charge state models, the charge ladder models, the Ohta–Kimura models just to mention a few names.

One way of studying the stepwise mutation model when the population obeys the laws of a general branching process may go through branching random walks and spatially homogeneous branching processes (cf. for example [Biggins, 1979] and [Laredo & Rouault, 1984]). Such an approach does however not fit within the scope of the present work in a natural manner. Many questions can however be asked about the charge levels of the ancestors of an RSI. If we for example let R_0 stand for the charge level of the RSI redefined as zero and let c_{-1}, c_{-2}, etc stand for the charge jumps ($c_{-k} \in \{-1, 0, +1\}$) of her successive ancestors, then $R_k =$ the charge level of $-k = \sum_{i=1}^k c_{-i}$. Clearly this defines a backward random walk on Z. Moreover, the sequence $(R_k, \tau(i_k, \omega_{-k}))_{k \in N}$ defines a (backward) Markov renewal process where the embedded Markov chain (with Z as state space) is either null recurrent or transient, according to whether

$$\hat{\mu}_{m^-}(\alpha) = \int_0^\infty e^{-\alpha t} E[\xi_{m^-}(dt)] = \hat{\mu}_{m^+}(\alpha) = \int_0^\infty e^{-\alpha t} E[\xi_{m^+}(dt)]$$

or not.

It may be of interest to study this process and other aspects of the model, specially in the critical case where some ideas in [Moran, 1975 & 1976] can probably be used. In the subcritical case the idea of reduced branching processes (cf. Fleishman et al, 1977) can be used to show that the charges of the individuals alive at some late time t conditionally upon non-extinction will not be far away from each other. Indeed this is due to the fact that such individuals will have a recent common ancestor.

Another interesting development is to investigate what happens when the mutation probability becomes very small. Here similar techniques as those used in the study of almost critical branching processes can be used (a recent reference where such processes are considered is [Jönsson, 1990]). A first investigation of this problem seems to be promising. Related references are [Fleishman, 1978] and [Sawyer, 1979].

Using multitype branching processes on general type spaces it is possible to construct a model for the balance between mutation and selection as in [Bürger, 1988]. Such a model can be described in many (more or less equivalent) ways. Consider for example the following set of assumptions:

- The reproduction point processes ξ, ξ_m and ξ_n are as above.

- The (geno-) type space Γ is a flexible set which can be Z, R^n etc.

- An individual of type γ in Γ has a fitness which expresses itself through a survival probability $s(\gamma)$ and can give birth to mutant offspring with types in the infinitesimal set $d\gamma'$ according to some probability density function $u(\gamma,\gamma')\, d\gamma'$.

Using some ideas in [Jagers, 1989] it is possible to obtain expressions for the stable distribution of types. One of the first problems one has to face is the following: Let

$$E_\gamma[\xi(d\gamma' \times du)] = \mu(\gamma, d\gamma' \times du)$$

denote the mean number of children with types in $d\gamma'$ born to an individual of type γ during the infinitesimal age interval du. The equation defining the Malthusian parameter is of the form

$$h(\gamma) = \int_\Gamma \int_{R_+} e^{-\alpha u}\, h(\gamma')\, \mu(\gamma, d\gamma' \times du)$$

(cf. the above reference) where h can be thought of as an eigen–function of the integral operator defining this equation. Using the above assumptions we get

$$h(\gamma) = s(\gamma)\, \mu_n(\hat{\alpha})\, h(\gamma) + s(\gamma)\, \mu_m(\hat{\alpha}) \int_\Gamma h(\gamma')\, u(\gamma,\gamma')\, d\gamma'.$$

If $\Gamma = \{1, 2, ..., k\}$ and

$$u(i,j) = u_j \quad (j \neq i)$$

$$u(i,i) = 1 - u - u_i$$

we will obtain a version of what Kingman (cf. [Kingman, 1980]) calls the house of cards model. In this simple case the stable distribution of types coincides with Kingman's. In the general case, however, it is not obvious that it will be possible to get any explicit results and one should investigate the existence of solutions to the above integral equation and thereby the existence of stable type distributions. In a second stage one should consider some of the special cases considered in the literature (cf. [Bürger, 1988]). In this context Kimura's (cf. [Kimura, 1965]) normal distribution result is of special interest.

BASIC CONVERGENCE THEOREMS

Consider a general branching process as described in Chapter One. Assume the following conditions are satisfied (processes satisfying these conditions are called well behaved)

(C.1) $\mu(x) = E[\xi(\infty)] > 1$

(C.2) $\exists\ \alpha$ such that $\hat{\mu}(\alpha) = \int_0^x e^{-\alpha u}\mu(du) = 1$

(C.3) $\beta = \int_0^x ue^{-\alpha u}\mu(du) < \infty$

(C.4) μ is non–lattice (i.e. cannot be supported by any lattice $\{0, \pm d, \pm 2d....\}, d > 0$).

For such processes the following results are valid

TH.1 ([Jagers & Nerman, 1984a])

Let $\chi \geq 0$ be a random characteristic meeting the following assumptions

(1.1) $\sum\limits_{n=0}^{\infty} \sup\limits_{n \leq u \leq n+1} e^{-\alpha u}E[\chi(u)] < \infty$

(1.2) $E[\chi]$ is continuous a.e.

Then

$$e^{-\alpha t}E[Z_t^{\chi}] \to E[\hat{\chi}(\alpha)]/\alpha\beta, \text{ as } t \to \infty.$$

TH. 2 ([Jagers & Nerman, 1984a])

Let χ be an individual random characteristic meeting (1.1)–(1.2) of TH.1 Assume further that

(2.1) $Var[\chi(t)]$ is bounded on finite intervals.

(2.2) $e^{-2\alpha t}Var[\chi(t)] \to 0$, as $t \to \infty$.

(2.3) $Var[\hat{\xi}(\alpha)] < \infty$.

Then

$$e^{-2\alpha t}Var[Z_t^{\chi}] \to \frac{E^2[\hat{\chi}(\alpha)]Var[\hat{\xi}(\alpha)]}{\alpha^2\beta^2(1-\hat{\mu}(2\alpha))}$$

as $t \to \infty$.

TH.3 ([Jagers & Nerman, 1984a])

If χ is a random characteristic such that (1.1) and (1.2) are satisfied and

(3.1) $\qquad E[\hat{\xi}(\alpha)\log^+\hat{\xi}(\alpha)] < \infty$

Then

$$\frac{Z_t^\chi}{y_t} \xrightarrow{\ P\ } E[\hat{\chi}(\alpha)] \quad \text{on } \{y_t \to \infty\}.$$

TH.4 ([Jagers & Nerman, 1984a])

Suppose that the random characteristic χ meets (1.1) and (1.2). Then

$$e^{-\alpha t}Z_t^\chi \xrightarrow{\ P\ } E[\hat{\chi}(\alpha)]W_\infty/\alpha\beta$$

on $\{y_t \to \infty\}$, as $t \to \infty$, with $W_\infty = \lim\limits_{t\to\infty} W_t = \sum\limits_{x\in I(t)} e^{-\alpha\sigma}x$ and

$$I(t) = \{xk \in I; \ \sigma_x \leq \sigma_{xk} < \infty\}, \ t \in R^+.$$

If in addition (3.1) is satisfied then the convergence holds also in L^1.

TH.5 ([Jagers & Nerman, 1984a])

Consider a random characteristic χ such that

(5.1) $\qquad \exists \ \alpha' < \alpha; \ E[\sup\limits_{t\geq 0} e^{-\alpha't}\chi(t)] < \infty.$

Assume further that the following condition is satisfied

(5.2) $\qquad \exists \alpha' < \alpha; \ \hat{\mu}(\alpha') = \int\limits_0^\infty e^{-\alpha't}\mu(dt) < \infty$

Then

$$\frac{Z_t^\chi}{y_t} \longrightarrow E[\hat{\chi}(\alpha)] \quad \text{a.s. on } \{y_t \to \infty\},$$

as $t \to \infty$.

TH.6 ([Nerman & Jagers, 1984)])

Assume $E[\hat{\xi}(\gamma)] < \infty$ for some $\gamma < \alpha$. Consider a measurable function $\varphi: \tilde{\Omega} \times R^+ \to R^+$, which has left and right limits in its second coordinate and satisfies

$$\sup\limits_{\tilde{\omega}, a} e^{-\gamma a}\varphi(\tilde{\omega}, a) < \infty.$$

Assume further that $\nu_\varphi(\tilde{\omega}, a)$ is such that $\tilde{P}(\nu_\varphi(\tilde{\omega}, a) < \infty) = 1$ and $\{a; \ \nu_\varphi(\tilde{\omega}, a) = n\}$, $n = 0,1,2,\ldots$ is the countable union of intervals whose endpoints are nowhere dense. Let φ_t be a stochastic process on $(\Omega^I, \mathscr{A}, P^I)$ obtained by summing $\varphi(\Pi_x, t-\sigma_x)$ for individuals x satisfying $\nu_\varphi(\Pi_x, t-\sigma_x) \leq n(x)$ and by some arbitrary bounded processes $\psi_x(t-\sigma_x)$ otherwise. Then

$$\varphi_t/y_t \to \tilde{E}[\varphi] \quad \text{a.s. on } \{y_t \to \infty\}$$

as $t \to \infty$.

The theorem below is formulated for branching processes but it holds more generally for (exponentially) growing stochastic processes. It relies upon a uniformity concept, which is named after Lebesgue: A class $\{f_t; t \geq 0\}$ of real valued functions of a real variable is Lebesgue conformable if (a) for any $T > 0$,

$$\sup_{t \geq 0, |u| \leq T} |f_t(u)| < \infty,$$

and (b) for any $\epsilon > 0$ the Lebesgue measure of

$$\{u; |u-u'|<\rho, |u-u''|<\rho \ |f_t(u') - f_t(u'')| > \epsilon\}$$

tends to zero uniformly in t, as $\rho \downarrow 0$. If (c) the series

$$\sum_{n=-\infty}^{+\infty} e^{-\alpha n} \sup_{n \leq u \leq n+1} |f_t(u)|$$

converges uniformly, for $\alpha > 0$, [Jagers & Nerman, 1984b] call the class α–Wiener.

These concepts are then applied to various functionals of stochastic processes $\{\chi_{t,x}(u); u \geq 0\}$, i.i.d. for fixed t, to be used as characteristics. E.g. consider for $\theta \in R$, χ_t generic for χ_{tx},

$$\varphi_{tu}(\theta) = E[e^{i\theta\chi_t(u)}],$$

$$h_t(u) = E[\chi_t(u)],$$

$$g_t(u) = E[|\chi_t(u)|^p],$$

and

$$D_{t\delta}(u) = \sup_{|v| < \delta} E[|\chi_t(u+v) - \chi_t(u)|^p],$$

for some $1 \leq p \leq 2$, all blown up by a factor $e^{\alpha t}$. For the last of these functions an average form of (b) is needed.

(b) $$\lim_{\delta \downarrow 0} \lim_{t \to \infty} \sup \int_{-\infty}^{+\infty} e^{\alpha t} D_{t\delta}(u)e^{-\alpha u} du = 0.$$

TH.7 ([Jagers & Nerman 1984b])

Assume given, for each $t \geq 0$, a set $\{\chi_{tx}(u)\}$ of random characteristics on the line such that $\{\xi_x, \chi_{tx}\}_{x \in I}$, are i.i.d. Denote the characteristic functions, expectations etc. as above and suppose that $\{e^{\alpha t} h_t; t \geq 0\}$ and $\{e^{\alpha t}(\varphi_{tu}(\theta) - 1); t \geq 0\}$, any fixed θ, are Lebesgue conformable and α–Wiener and further that, for some $p, 1 \leq p \leq 2$, and all small $\delta > 0$, $g_t(u)$ and $D_{t\delta}(u)$ also define Lebesgue conformable classes $\{e^{\alpha t} g_t\}$ and $\{e^{\alpha t} D_{t\delta}\}$, δ fixed.

Assume that

$$\lambda = \lim_{t \to \infty} \alpha \int_{-\infty}^{+\infty} e^{\alpha t} h_t(u) e^{-\alpha u} du \text{ and}$$

$$\hat{\psi}_\alpha(\theta) = \lim_{t \to \infty} \int_{-\infty}^{+\infty} e^{\alpha t} \{\varphi_{tu}(\theta) - 1\} e^{-\alpha u} du$$

exist and are finite and that (b) holds.

Let W_∞ denote the limit in probability of the normed total population

$$e^{-\alpha t} y_t \xrightarrow{P} W_\infty / \alpha\beta.$$

Then

$$Z_t^{\chi_t} = \sum_{x \in I} \chi_{tx}(t - \sigma_x)$$

converges in distribution, as $t \to \infty$, to a random variable whose characteristic function is

$$E[\exp(W_\infty \hat{\psi}_\alpha(\theta)/\alpha\beta)].$$

THE LATTICE CASE

Consider a general branching process in the sense of Chapter One and assume that the reproduction measure μ is concentrated on some lattice $\{0, d, 2d,...\}$ for some $d \in N$. for simplicity we shall take $d=1$. The aim of this section is to formulate counterparts of some of the basic convergence theorems stated in APPENDIX A. The proofs will follow mutatis mutandis from those in [Jagers & Nerman, 1984a] for the supercritical case.

The Malthusian parameter, α, will now be given by

$$\sum_{j=0}^{\infty} e^{-\alpha j} \mu(\{j\}) = 1.$$

If we now assume that everything is defined for discrete time, we can introduce τ_k, $\tau(k,\omega)$, $\xi(k)$, $\chi(k)$, Z_k^χ, y_k, m_k^χ etc in obvious notation, $k \in Z^+$.

The following two results are consequences of the lattice version of the key renewal theorem [Jagers, 1975] and some calculations

TH.1

$$\lim_{k \to \infty} e^{-\alpha k} m_k^\chi = \sum_j e^{-\alpha j} E[\chi(j)] / \sum_j j e^{-\alpha j} \mu(\{j\})$$

where we assume $\sum_j e^{-\alpha j} E[\chi(j)] < \infty$ and the same for $\beta = \sum_j j e^{-\alpha j} \mu(\{j\})$ which is of course recognized as the mean age at childbearing in the lattice case.

TH.2

Assume that χ is of the individual type and that the conditions of TH.1 are satisfied. If further

      (i) $Var[\chi(k)]$ is bounded

      (ii) $\lim_{k \to \infty} e^{2\alpha k} Var[\chi(k)] = 0$

      (iii) $Var[\sum_j e^{-\alpha j} \xi(\{j\})] < \infty.$

Then

$$\lim_{k\to\infty} e^{-2\alpha k} \mathrm{Var}[Z_k^\chi] = \left[\frac{\sum_j e^{-\alpha j} E[\chi(j)]}{\sum_j j e^{-\alpha j} \mu(\{j\})}\right]^2 \frac{\mathrm{Var}[\sum_j e^{-\alpha j}\xi(\{j\})]}{1 - \sum_j e^{-2\alpha j}\mu(\{j\})} \cdot$$

The following two results are direct translations to the lattice case of TH.4 and TH.5 in APPENDIX A.

TH.3

Under the conditions of TH.1

$$e^{-\alpha k} Z_k^\chi \xrightarrow{\ P\ } \sum_j e^{-\alpha j} E[\chi(j)]W_\infty / \sum_j j e^{-\alpha j}\mu(j)$$

on $\{y_k \to \infty\}$, as $k \to \infty$ with

$$W_\infty = \lim_{k\to\infty} W_k \quad \text{and} \quad W_k = \sum_{x\in I(k)} e^{-\alpha\sigma_x}$$

where

$$I(k) = \{xi;\ x \in I,\ i\in N \text{ and } \sigma_x \le k < \sigma_{xi} < \infty\},$$

$K \in Z_+$. If in addition

(3.1) $E[\sum_j e^{-\alpha j}\xi(j)\log^+(\sum_j e^{-\alpha j}\xi(j))] < \infty$

the convergence holds also in L^1.

TH.4

Assume that condition (3.1) in TH.3 holds and that $\sum_j e^{-\alpha j} E[\chi(j)] < \infty$. Then as

$k \to \infty$

$$Z_k^\chi / y_k \xrightarrow{\ P\ } (1-e^{-\alpha}) \sum_j e^{-\alpha j} E[\chi(j)]$$

on $\{y_k \to \infty\}$.

A LIST OF SYMBOLS AND CONVENTIONS

The following list contains symbols and conventions strictly adhered to throughouts the main text and does not include symbols and conventions restricted to one chapter.

LATIN ALPHABET

a	Stands mostly for age
$\mathscr{B}$	The Borel algebra
$\xi^{(n)}(t)$	The number of individuals in the nth generation born before t
$E[X;A]$	$= E[X1_A]$
$\tilde{E}$	Expectation with respect to the stable pedigree measure
i_k	The birth order of $-(k-1)$ among the offspring of $-k$
I	The set of all individuals
$-j$	The jth ancestor of the RSI
J	$(-Z^+) \times I$
M_t	The number of living labels at t
$n(x)$	The generation number of the individual x
N	The positive integers
N^n	The nth generation
N_t	The total number of labels at t
P	The life law
$\tilde{P}$	The stable pedigree measure
q_n	The probability of ultimate extinction of the Ancestral process.
$q_n(t)$	The probability of extinction before t of the Ancestral process
R	$(-\infty, \infty)$
R_+	$[0, \infty)$
S_x	The shift operator $\{\omega_y; y \in I\} \xrightarrow{S_x} \{\omega_{xy}; y \in I\}$
t,u,v	Denote usually time
W_∞	The limit of $\alpha\beta e^{-\alpha t} y_t$
x	Individual

x_i	The ith coordinate of x
$x_{[k]}$	The kth last ancestor of $x = (x_1,...,x_{n(x)-k})$, $k \leq n(x)$
y	Individual
y_t	The total population process
$y_n(t)$	The total population process corresponding to the Ancestral process
$Z^+ \; Z^-$	The non–negative and the negative integers respectively
Z_t^n	The process counting living individuals at t in the Ancestral process
Z_t^χ	The branching process counted by the random characteristic χ

GREEK ALPHABET

α	The malhtusian parameter
β	The mean age at childbearing $\left(= \int_0^\infty t e^{-\alpha t} \mu(dt)\right)$
λ	The life span
μ	The reproduction measure $(= E[\xi])$
μ_m	$= E[\xi_m]$
$\mu_\alpha^m(t)$	$= \int_0^t e^{-\alpha u} \mu_m(du)$
ν	$= \sum_{k=0}^\infty \mu^{*k}(t) = E[y_t]$
ν_n	$= E[y_n(t)]$
ξ	The reproduction point process
$\xi(t)$	$= \xi([0,t])$
ξ_m	The reproduction point process of mutant children
ξ_n	The reproduction point process of non–mutant children
$\xi_{m,x}$	The ξ_m – process pertaining to x
$\rho(i,\omega_x)$	The mutation index of the ith child of x
σ_x	The birth time moment of the individual x

$\tilde{\sigma}_x$	The birth time moment of the individual x in the notation of the stable pedigree calculus
$\tau(k,\omega_x)$	The age at which x begets her kth child
$\tau(k)$	The same as $\tau(k,\omega_0)$
φ	Denotes usually some functional of the pedigree of the RSI
χ	A random characteristic
ω	The individual life carreer or an element of Ω^I
Ω	The set of all permissible life carreers
$\tilde{\Omega}$	$= \Omega^J \times N^\infty$
$\tilde{\omega}$	An element of $\tilde{\Omega}$

SYMBOLS

$\xrightarrow{\ p\ }$	Convergence in probability
$\longrightarrow$	Maps or converges
#	Cardinality
*	Convolution $(f*g(t) = \int_0^t f(t-u)g(du))$
1_A	The indicator function of the set A
2^I	The class of all subsets of I
>	x > y means that x stems from y (x,y individuals)
$\square$	Marks the end of a proof
$\sim$	Asymptotically equals
$\char94$	Denotes the Laplace–Stiltjes transform

ABBREVIATIONS

a.e.	Almost everywhere
a.s.	Almost surely
d.f.	Distribution function
i.i.d.	Independent and identically distributed
p.g.f.	Probability generating function
r.v.	Random variable
RSI	Randomly samples individual
w.p.	With probability
w.r.t.	With respect to
"x logx"	By the x logx–condition is meant the requirement that $E[\hat{\xi}(\alpha) \log \hat{\xi}(\alpha)] < \infty$

REFERENCES

ABRAMOWITZ M. & STEGUN I.A. (1965) Handbook of Mathematical Functions. Dover Publ. Inc. New York.

ASMUSSEN S. & HERING H. (1983) Branching Processes. Birkhauser, Boston.

ATHREYA K.B. & NEY P.E. (1972) Branching Processes. Springer Verlag. Berlin Heidelberg New York.

BIGGINS J.D. (1979) Growth rates in the branching random walk. Z. Wahrscheinlichkeitsteorie Verw. Gebeite 48, 17-34.

BROBERG P. (1987) Sibling dependencies in branching populations. (Thesis) Dept. of Math. University of Göteborg, Sweden.

BÜHLER W. (1971) Generations and the degree of relationship in a supercritical Markov branching process. Z. Wahrsheinlichkeitsteorie Verw. Gebiete. 18, 141-152.

BÜHLER W. (1972) The distribution of generations and other aspects of the family structure of branching processes. In Proc. 6th Berkeley Symp. Math. Statist. Prob. 3, 463-480.

BÜRGER R. (1988) Mutation-selection balance and continuum of alleles models (Unpublished).

CHAUVIN B. (1986) Arbres et processus de Bellman-Harris. Ann. Inst. Henri Poincaré. Vol. 22, n° 2, 209-232.

CHAKRABORTY R. (1975) Nucleotide differences between two randomly chosen cistrons in a population of variable size. Theor. Pop. Biol. 11, 11-22.

DONEY R.A. (1976) The single and multitype general age dependent branching processes. J. Appl. Prob. 13, 239-246.

EWENS W.J. (1978) The neutralist-selectionist controversy. Math. Ass. of America 16, 477-502.

EWENS W.J. (1979) Mathematical Population Genetics. Springer Verlag, Berlin.

EWENS W. (1990) Population genetics theory - the past and the future. In Proc. NATO ASI Series. 299, 177-227.

FISHER R.A. (1922) On the dominance ratio. Proc. Royal Soc. Edinburgh. 42, 321-341.

FLEISHMAN K. & SIEGMUND-SCHULTZE U.R. (1977) The structure of reduced critical Galton-Watson processes. Math. Nachr. 79, 233-241.

FLEISHMAN J. (1978) Limiting distributions for critical branching Brownian random fields. Trans. Amer. Math. Soc. 239, 353-389.

HOPPE F.M. (1984) Limit behaviour of an urn model in population genetics. (Thechnical report). Dept. of Statistics. University of Michigan.

JACQUARD A. (1974) The Genetic Structure of Populations. Springer-Verlag, New York.

JAGERS P. (1974) Convergence of general branching processes and functionals thereof. J. Appl. Prob. 11, 471-478.

JAGERS P. (1975) Branching Processes with Biological Applications. Wiley, Chichester.

JAGERS P. (1982) How probable is it to be first born? and other branching process applications to kinship problems. Math. Biosciences 59, 1-15.

JAGERS P. (1989) General branching processes as Markov fields. Stoch. Proc. Appl. 32, 182-212.

JÖNSSON T. (1990) Diffusion approximation and extinction of almost critical general branching processes. (Preprint) Dept. of Math. University of Göteborg, Sweden.

JAGERS P. & NERMAN O. (1984a) The growth and composition of branching populations. Adv. Appl. Prob. 16, 221-259.

JAGERS P. & NERMAN O. (1984b) Limit theorems for sums determined by branching processes and other exponentially growing processes. Stoch. Proc. Appl. 17, 47-71.

KALLENBERG O. (1983) Random Measures, 3rd edition. Akademic-Verlag, Berlin/Academic Press, New York.

KARLIN S. & McGREGOR J. (1967) The number of mutant forms maintained in a population. In Proc. 5th Berkeley Symp. Math. Statist. Prob. 4, 415-438.

KIMURA M. (1965) A stochastic model concerning the maintenance of genetic variability in quantitative characters. Proc. Natl. Acad. Sci. USA 54, 731-736.

KIMURA M. (1968) Evolutionary rate at the molecular level. Nature 217, 624-662.

KIMURA M. (1969) The number of heterozygous nucleotide sites maintained in a finite population due to a steady flux of mutations. Genetics 61, 893.

KIMURA M. (1971) Theoretical foundations of population genetics at the molecular level. Theor. Pop. Biol. 2, 174-208.

KIMURA M. & CROW J.F. (1964) The number of alleles that can be maintained in a finite population. Genetics 49, 725-738.

KIMURA M. & OHTA T. (1973) A model of mutation appropriate to estimate the number of electrophoretically detectable alleles in a finite population. Genet. Res. Cambr. 22, 201-204.

KIMURA M. (1983) The Neutral Theory of Molecular Evolution. Cambridge University Press, Cambridge.

KING J.L. & JUKES T.H. (1969) Non-Darwinian theory of evolution: Random fixation of a selectively neutral mutation. Science 164, 788-798.

KINGMAN J.F.C. (1980) Mathematics of Genetic Diversity. SIAM. Philadelphia, Pennsylvania.

LAREDO C. & ROUAULT A. (1983) Grandes déviations, dynamique de populations et phénomènes Malthusiens. Ann. Inst. Henri Poincaré, Vol. XIX, 4, 323-350.

LEADBETTER M.R., LINDGREN G. & ROOTZÉN H. (1983) Extremes and Related Properties of Random Sequences and Processes. Springer-Verlag, New York Heidelberg Berlin.

MORAN P.A.P. (1975) Wandering distributions and the electrophoretic profile. I. Theor. Pop. Biol. 8, 318-330.

MORAN P.A.P. (1976) Wandering distributions and the electrophoretic profile. II. Theor. Pop. Biol. 10, 145-149.

NERMAN O. (1981) The convergence of supercritical general (C-M-J) branching processes. Z. Wahrscheinlichkeitsteorie Verw. Gebiete 57, 365-395.

NERMAN O. (1984a) The stable pedigrees of critical branching processes. J. Appl. Prob. 21, 447-463.

NERMAN O. (1984b) The growth and composition of supercritical branching populations on general type spaces. (Preprint) Dept. of Mathematics. University of Göteborg, Sweden.

NERMAN O. (1987) Branching processes and population genetics. In Proceedings of the first Bernoulli Congress held in Tashkent. V N U Science Press. Utrecht, Holland.

NERMAN O. & JAGERS P. (1984) The stable doubly infinite pedigree process of supercritical branching processes. Z. Wahrscheinlichkeitstheorie Verw. Gebiete 65, 445-460.

NEVEU J. (1986) Arbres et processus de Galton-Watson. Ann. Inst. Henri Poincaré. Vol. 22, n° 2, 199-207.

PAKES T. (1984) Coloured branching processes. (Unpublished.)

PAKES T. (1987) An infinite alleles version of the Markov branching process (Personal communication).

RYAN T.A. Jr. (1968) On Age-dependent branching processes (Thesis) Cornell University.

ROSS S. (1982) Stochastic Processes - Wiley, New York.

SAWYER S. (1975) An application of branching random fields to genetics. In Probablistic Methods in Differential Equations. Springer-Verlag. Lecture Notes in Mathematics.

SAWYER S. (1976) Branching diffusion processes in population genetics. Adv. Appl. Prob. 8, 659-689.

SAWYER S. (1979) A limit theorem for patch sizes in a selectively neutral migration model. J. Appl. Prob. 16, 482-495.

STEBBINS G.L. & AYALA F.J. (1985) The evolution of Darwinism. Scientific American, July.

TAVARÉ S. (1984) Lines of descent and genealogical processes and their applications in population genetics. Theor. Pop. Biol. 26, 119-164.

TAVARÉ S. (1989) The genealogy of birth, death and immigration processes. In Mathematical Evolutionary Theory, ed. M.W. Feldman, 41-56. Princeton University Press, Princeton.

WALLACE B. (1981) Basic Population Genetics. Columbia Univ. Press, New York.

WATTERSON G.A. (1976) Reversibility and the age of an allele I. Theor. Pop. Biol. 10, 239-253.